Engineering Elasticity

Humphrey Hardy

Engineering Elasticity

Elasticity with less Stress and Strain

 Springer

Humphrey Hardy
Pelzer, SC, USA

The book has solutions to the exercises that are available to Professors only. These can be viewed at: https://link.springer.com/book/10.1007/978-3-031-09157-5.

ISBN 978-3-031-09159-9   ISBN 978-3-031-09157-5   (eBook)
https://doi.org/10.1007/978-3-031-09157-5

This Springer imprint is published by the registered company Springer Nature Switzerland AG
The registered company address is: Gewerbestrasse 11, 6330 Cham, Switzerland

Preface for Students

Historically, the materials available to engineers for construction consisted of materials that fail when deformed by relatively small amounts (e.g., stone, wood, and concrete). To build computer simulations for applications of these materials required simulating only small deformations. This has been done quite successfully using finite elements and matrices. This approach is called linear elasticity.

The advent of materials which can be deformed by large amounts without failing or fracturing (e.g., rubber, plastics, and biological materials) requires a different approach. A new approach is needed because large bending, compression, or extension (finite deformations) results in equations of motion that are nonlinear. To address the problem of finite deformations, classical elasticity has introduced many measures of deformation (strain) and the forces to produce the deformations (stress). Stress and strain can be defined in terms of matrices (second-order tensors), and classical elasticity texts relate stress and strain using higher-order tensor mathematics. This makes the mathematics of large deformations difficult, requiring graduate courses in stress, strain, and tensor analysis.

A different approach is taken in this text. Only one measure of stress and one measure of strain is defined. The relationship between these two measures is defined in terms of energy instead of a direct relationship between stress and strain. The result is that many finite deformations can be simulated with only matrices (no higher-order tensors needed). Thus, all that is required is a familiarity with calculus-based physics and linear algebra.

The mathematics needed, example computer simulations, and experiments to measure material properties for large deformations of materials are all included here. The mathematics begins with a brief review of the physics and linear algebra. Deformations are described in terms of mapping between the initial and final states of the material. Forces and the relationship between force and energy follow. The constraints required on the energy function for isotropic materials are described. Next, simulations are shown using Mathematica. These are examples which test and extend the theory. A very inexpensive experimental set up is described to show how material properties can be measured. Time-dependent simulations are described, as well as the application of this approach to anisotropic materials. A number of different types of output plots are presented.

A more eloquent derivation of the equations is made by using the techniques of Euler and Lagrange, but this requires higher-level physics and mathematics. A chapter is included to show how the equations of linear elasticity follow from the finite elasticity equations. Finally, a chapter is included for those students already familiar with classical elasticity to compare the approach of using a single stress and strain measure with classical elasticity descriptions.

Pelzer, SC, USA   Humphrey Hardy

Preface for Instructors

This book is an engineering approach to finite deformations. It is not intended to be an introduction to classical finite elasticity. Classical finite elasticity introduces a dozen or more stress and strain measures, makes use of the tensor property of these measures, and expresses the equation of motion in terms of both the reference and body coordinates. This approach does none of these things. To see the connection of this approach with classical finite elasticity, see Chap. 16.

The approach here is to introduce the engineer to only what is needed to solve a broad range of deformations. The approach is based on the same model that is used in teaching engineers linear elastic deformations. In linear elasticity, engineers are introduced to Cauchy stress and strain and none of the many other measures of classical finite elasticity. Cauchy-Green tensors, Kirchhoff stress, and Lagrangian strain are not introduced in linear elasticity because they are not needed by the engineer to simulate applications of small deformations and small rotations. In this text, the engineer is introduced to only one stress and one strain measure, which is all that is needed to solve many finite deformation problems. Because of this simplification, the target audience for this approach can be second-year or third-year undergraduates, but as discussed below, with a slightly different order of chapters, this text could also be used for seniors or graduate students.

The relationship between any finite stress and strain measure is expressed in terms of non-linear differential equations and therefore requires numerical simulations for applications. The body of the text uses a minimization technique using Mathematica to numerically solve a number of examples. There are many other techniques and software available which could be used to solve these examples, but this is not a text on writing computer code. Instead, the examples simply show some of the breath of what is possible with a single stress and strain measure. The codes used are found online with comments so that the enterprising student can read and translate this code into any language they wish. No homework problems require Mathematica; however, an algebraic software or computer graphics program would be useful for some of the homework problems. When these are needed, the individual problems state this.

Engineers also require experimental measurements of materials to provide the material property parameters that occur in the differential equations of elasticity. Chapter 10 is included to illustrate one method of making physical measurements on a material and turning these measurements into the needed material parameters. The experiment is simple and cost effective so that it can be included in any course. The enterprising student should have no problem either reproducing the experiments or designing more accurate experiments using more sophisticated (and expensive) tools if this is of interest to the instructor and student. In short, this text provides all an engineer needs to model many materials in finite deformations.

Any engineer that requires the description of large deformations can benefit from this course. Materials like rubber, soft plastics, and biological materials often undergo large deformations when placed in service. Materials like spaghetti, cake icing, and car bumpers are often extruded and also undergo large deformations. Even engineers producing movies can benefit from a correct description of large deformations of materials. This book provides theory, experimental measurements of properties, and numerical simulations of these types of large deformations. In each chapter of the book, 8–12 problems have been provided. Solutions have been provided in the solutions manual. Mathematica software has been provided (online) for the examples and figures in the text. The software is not designed to be commercial code but is provided so that no steps in the simulation are omitted and the code can be used as a model for the student to either extend the models or translate them into other computer languages. The Mathematica code can be read using the free Mathematica player found at https:// wolfram.com/player/, but to change or execute the code, a Mathematica license, found at https:\\wolfram.com\mathematica\, is required.

A basic undergraduate course would need to cover Chaps. 1, 2, 3, 4, 5, and 6. After that, the course can go in many directions depending upon the interest of the student and professor. For applications of anisotropic materials, Chaps. 9 and 12 should be added. For the student wishing to develop software tools, Chaps. 7, 8, 11, and 13 would be helpful, but familiarity with programming would be needed for these chapters. Chapter 10 covers experimental measurements of material properties. Chapter 14 would be appropriate for more advanced senior or graduate students familiar with Lagrangians. For more advanced students, Chaps. 1, 2, 3, 4, 5, and 6 could be omitted and the applications in Chap. 7 and onward approached immediately. Chapters 15 and 16 are included to make the connection of this approach to linear elasticity and classical finite elasticity and would also be appropriate for senior undergraduate or graduate students.

Appendix A derives the mapping between the fixed coordinate system of the simulation and the experimental coordinate system used to define anisotropy properties. Appendix B provides a description of Gram-Schmidt QRD which is the basis of the anisotropic invariants found in Chap. 12. Appendix C provides a quick review of Euler-Lagrange equations in multiple dimensions for those students with some familiarity with Lagrangians. Appendix D provides the code of an example showing that finite element techniques also solve the equations of finite elasticity. Appendix E provides a list 50 "projects", each of which would be appropriate for a class assignment or an undergraduate or master's thesis depending upon the project. These may be used "as is" or as a starting point for a student's own project ideas.

Pelzer, SC, USA   Humphrey Hardy

Contents

About the Author

Humphrey Hardy holds a BS from Louisiana Tech University, an MS from the University of Maryland, and a PhD from the University of Houston. He has taught engineers undergraduate physics for 19 years. He also worked full time in R&D in the oil industry in reservoir engineering, seismic processing, and faulting and fracturing for 24 years. He is the author of the book, *Fractals in Reservoir Engineering*. He has been a distinguished lecturer for the Society of Petroleum Engineering and the Chemical Research Council. He holds six US Patents.

Chapter 1
Getting Ready (Mostly Review)

This chapter reviews the basic equations of linear algebra and Newtonian physics that will be needed in future chapters. A section entitled "Miscellaneous" is included to cover some things that may not be common knowledge for all students.

Linear Algebra

Scalars, Vector, and Matrices

A **scalar** is a quantity which describes a property with a single value.

Example scalars: 5, 7 seconds, a, b, etc.

A **vector** is a pair of numbers in two dimensions and a set of three numbers in three dimensions to describe a single property. The individual numbers are called the components of the vectors. Vectors are often represented as an arrow. When arrows are used, the length and direction of the arrow are defined by the components of the vector, but the tail of the arrow is left to be defined as needed.

Example vectors in 2D: (2, 3), (4 m, 2 m), (a, b), etc.

Example vectors in 3D: (5, 4, 7), (a, b, c), (2m, 1m, 7m).

Some authors use $\Leftrightarrow$ brackets for vectors and () brackets for points. The different brackets are useful because they can distinguish between a vector $\vec{v}$ and a point $\vec{p}$; however, in this text, () brackets will be used for both.

A vector can be represented as a single letter by placing an arrow over the letter.

Example vector notations: $\vec{a}$, $\vec{v}$, $\vec{s}$ (e.g., $\vec{a}$ may stand for (2, 3, 7)).

The components of a vector can be represented with x, y, and z subscripts or 1, 2, and 3 subscripts:

H. Hardy, *Engineering Elasticity*, https://doi.org/10.1007/978-3-031-09157-5_1

$$\vec{a} = (a_x, a_y, a_z) \quad \text{or} \quad \vec{a} = (a_1, a_2, a_3). \tag{1.1}$$

Note that the vector letter $\vec{a}$ has an arrow over it to show that it is a vector, but the components of a vector, for example, a_x or a_1, are scalars and hence do not have arrows over them.

Matrices are defined as a set of four numbers in 2D and nine numbers in 3D. They are written in rows and columns, e.g., $\begin{pmatrix} 1 & 3 \\ 7 & 5 \end{pmatrix}$ or $\begin{pmatrix} 1 & 0 & 2 \\ 2 & 4 & 3 \\ 6 & 1 & 5 \end{pmatrix}$. They will

be represented in this book by script letters, e.g., $\mathcal{F}$ or $\mathcal{T}$, etc. The individual numbers that make up a matrix are called the elements of the matrix. A matrix can also be represented as follows:

$$\mathcal{F} = \begin{pmatrix} \mathcal{F}_{11} & \mathcal{F}_{12} & \mathcal{F}_{13} \\ \mathcal{F}_{21} & \mathcal{F}_{22} & \mathcal{F}_{23} \\ \mathcal{F}_{31} & \mathcal{F}_{32} & \mathcal{F}_{33} \end{pmatrix} \tag{1.2}$$

Notice carefully the order of the subscripts. The second subscript is incremented as you move from column to column across the array; the first subscript is incremented as you move row to row down the array.

The three column vectors, $\vec{a}$, $\vec{b}$, and $\vec{c}$, which make up $\mathcal{F}$ are defined as

$$\begin{aligned} \vec{a} &= (\mathcal{F}_{11}, \mathcal{F}_{21}, \mathcal{F}_{31}) \\ \vec{b} &= (\mathcal{F}_{12}, \mathcal{F}_{22}, \mathcal{F}_{32}) \\ \vec{c} &= (\mathcal{F}_{13}, \mathcal{F}_{23}, \mathcal{F}_{33}). \end{aligned} \tag{1.3}$$

The three row vectors which make up $\mathcal{F}$ are defined as

$$\begin{aligned} \vec{m} &= (\mathcal{F}_{11}, \mathcal{F}_{12}, \mathcal{F}_{13}) \\ \vec{n} &= (\mathcal{F}_{21}, \mathcal{F}_{22}, \mathcal{F}_{23}) \\ \vec{s} &= (\mathcal{F}_{31}, \mathcal{F}_{32}, \mathcal{F}_{33}) \end{aligned} \tag{1.4}$$

An element of $\mathcal{F}$ may be represented as $\mathcal{F}_{ij}$, where i and j may take on the values of 1 and 2, respectively, in two dimensions and the values of 1, 2, and 3 in three dimensions. Note that a letter with i, j subscripts is an element of a matrix. The same letter without a subscript or with a single subscript is still a matrix.

For example, $\mathcal{F}_{jig}$ and $\mathcal{F}$ represent matrices. $\mathcal{F}_{ij}$ represents an element of the matrix $\mathcal{F}$.

Addition and Subtraction of Vectors

With $\vec{a} = (a_1, a_2, a_3)$ and $\vec{b} = (b_1, b_2, b_3)$

$$\vec{a} + \vec{b} = (a_1 + b_1, a_2 + b_2, a_3 + b_3), \tag{1.5}$$

So, if $\vec{a} = (1, 2, 3)$ and $\vec{b} = (1, 0, 1)$, then

$$\vec{a} + \vec{b} = (2, 2, 4). \tag{1.6}$$

In Einstein notation $\vec{a} + \vec{b}$ can be written as

$$\left(\vec{a} + \vec{b} \right)_i = a_i + b_i \tag{1.7}$$

for $i = 1, 2,$ and 3 in three dimensions and $i = 1$ and 2 for two dimensions.
Vectors are subtracted in the same manner:

$$\left(\vec{a} - \vec{b} \right)_i = a_i - b_i. \tag{1.8}$$

Multiplication of a Scalar Times a Vector

A scalar times a vector is

$$a \vec{v} = a(v_x, v_y, v_z) = (av_x, av_y, av_z). \tag{1.9}$$

For example, if $\vec{v} = (2, 3, 6)$ and $a = 2$,

$$2 \vec{v} = (4, 6, 12). \tag{1.10}$$

In Einstein notation the multiplication of a times $\vec{v}$ is

$$\left(a \vec{v} \right)_i = a v_i \tag{1.11}$$

for $i = 1, 2, 3$ or $i = 1, 2$.

Multiplication of Vectors

There are two ways to multiply vectors. One is called the dot product and the other is called the cross product. They are defined and designated as follows:

Dot product:

$$\vec{a} \circ \vec{b} = a_1 b_1 + a_2 b_2 + a_3 b_3. \tag{1.12}$$

In Einstein notation,

$$\vec{a} \circ \vec{b} = a_i b_i, \tag{1.13}$$

where it is assumed in Einstein notation that repeated indices in products are summed over 1, 2, and 3 in 3D and over 1 and 2 in 2D.

Cross product (only defined in 3D):

$$\vec{a} \times \vec{b} = (a_2 b_3 - a_3 b_2, a_3 b_1 - a_1 b_3, a_1 b_2 - a_2 b_1). \tag{1.14}$$

A physical interpretation of $\vec{a} \times \vec{b}$ can be made if the tails of the $\vec{a}$ and $\vec{b}$ vectors are placed at the same point. Then the area of the triangle defined by $\vec{a}$ and $\vec{b}$ is

$$\text{Area} = 1/2 \sqrt{\left(\vec{a} \times \vec{b}\right) \circ \left(\vec{a} \times \vec{b}\right)}. \tag{1.15}$$

Another useful product is the triple scalar product, defined as

$$\vec{a} \circ \left(\vec{b} \times \vec{c}\right) = a_1(b_2 c_3 - b_3 c_2) + a_2(b_3 c_1 - b_1 c_3) + a_3(b_1 c_2 - b_2 c_1). \tag{1.16}$$

This is the same as the determinant of the matrix

$$\left| \begin{pmatrix} a_1 & b_1 & c_1 \\ a_2 & b_2 & c_2 \\ a_3 & b_3 & c_3 \end{pmatrix} \right| = \vec{a} \circ \left(\vec{b} \times \vec{c}\right), \tag{1.17}$$

where

$$\vec{a} = (a_1, a_2, a_3)$$
$$\vec{b} = (b_1, b_2, b_3)$$
$$\vec{c} = (c_1, c_2, c_3).$$

If the tails of $\vec{a}$, $\vec{b}$, and $\vec{c}$ are located at the same point, the magnitude of the triple scalar product has the physical interpretation of six times the volume of the tetrahedron formed by these three vectors, i.e.,

$$\text{volume} = 1/6 \mid \vec{a} \circ \left(\vec{b} \times \vec{c} \right) \mid . \tag{1.18}$$

Addition and Subtraction of Matrices

Matrices must have the same number of elements to add or subtract them. You cannot add a 2D matrix to a 3D matrix. When you add matrices, you add the corresponding element of each matrix to obtain a new matrix. Thus

$$C = A + B, \tag{1.19}$$

is

$$C = \begin{pmatrix} a_{11} & b_{12} & c_{13} \\ a_{21} & b_{22} & c_{23} \\ a_{31} & b_{32} & c_{33} \end{pmatrix} + \begin{pmatrix} b_{11} & b_{12} & b_{13} \\ b_{21} & b_{22} & b_{23} \\ b_{31} & b_{32} & b_{33} \end{pmatrix}$$

$$= \begin{pmatrix} a_{11} + b_{11} & a_{12} + b_{12} & a_{13} + b_{13} \\ a_{21} + b_{21} & a_{22} + b_{22} & a_{23} + b_{23} \\ a_{31} + b_{31} & a_{32} + b_{32} & a_{33} + b_{33} \end{pmatrix} \tag{1.20}$$

In Einstein notation the elements of C, c_{ij}, can be written as $a_{ij} + b_{ij}$, where a_{ij} and b_{ij} are the elements of matrix A and B, respectively. That is,

$$c_{ij} = a_{ij} + b_{ij}. \tag{1.21}$$

Matrices are subtracted in the same manner:

$$(A - B)_{ij} = a_{ij} - b_{ij}. \tag{1.22}$$

A Scalar Times a Matrix

A scalar times a matrix multiplies every element of the matrix:

$$(a\,B)_{ij} = a\,b_{ij} \tag{1.23}$$

Multiplication of Matrices

Matrix multiplication is defined as each row of the first matrix times each column of
the second matrix. It is expressed most compactly in the Einstein notation. The i, j
element of $\mathcal{A}$ times $\mathcal{B}$ is

$$(\mathcal{A}\,\mathcal{B})_{ij} = a_{ik}\,b_{kj}. \tag{1.24}$$

Since repeated indices are summed over, in 3D Eq. 1.24 is also

$$(\mathcal{A}\,\mathcal{B})_{ij} = \sum_{k=1}^{3} a_{ik}\,b_{kj} \tag{1.25}$$

so that the $i = 1, j = 1$ element of $\mathcal{A}$ times $\mathcal{B}$ is

$$(\mathcal{A}\,\mathcal{B})_{1,1} = a_{11}b_{11} + a_{12}\,b_{21} + a_{13}\,b_{31} \tag{1.26}$$

The $i = 1, j = 2$ element is

$$(\mathcal{A}\,\mathcal{B})_{1,2} = a_{11}\,b_{12} + a_{12}\,b_{22} + a_{13}\,b_{32} \tag{1.27}$$

etc.

In full form,

$$
\mathcal{A}\,\mathcal{B} = \begin{pmatrix} a_{11} & a_{12} & a_{13} \\ a_{21} & a_{22} & a_{23} \\ a_{31} & a_{32} & a_{33} \end{pmatrix} \begin{pmatrix} b_{11} & b_{12} & b_{13} \\ b_{21} & b_{22} & b_{23} \\ b_{31} & b_{32} & b_{33} \end{pmatrix}
$$

$$
= \begin{pmatrix} a_{11}b_{11} + a_{12}b_{21} + a_{13}b_{31} & a_{11}b_{12} + a_{12}b_{22} + a_{13}b_{32} & a_{11}b_{13} + a_{12}b_{23} + a_{13}b_{33} \\ a_{21}b_{11} + a_{22}b_{21} + a_{23}b_{31} & a_{21}b_{12} + a_{22}b_{22} + a_{23}b_{32} & a_{21}b_{13} + a_{22}b_{23} + a_{23}b_{33} \\ a_{31}b_{11} + a_{32}b_{21} + a_{33}b_{31} & a_{31}b_{12} + a_{32}b_{22} + a_{33}b_{32} & a_{31}b_{13} + a_{32}b_{23} + a_{33}b_{33} \end{pmatrix}
$$

$$\tag{1.28}$$

One peculiar thing about multiplication of matrices is that the order of the matrices in
multiplication matters. That is

$$\mathcal{A}B \neq B\mathcal{A}. \tag{1.29}$$

Multiplication of a Matrix Times a Vector

The multiplication of a matrix times a vector is defined as each row of the matrix
times the components of the vector expressed in a column:

$$\mathcal{A}\,\vec{v} = \begin{pmatrix} a_{11} & a_{12} & a_{13} \\ a_{21} & a_{22} & a_{23} \\ a_{31} & a_{32} & a_{33} \end{pmatrix} \begin{pmatrix} v_1 \\ v_2 \\ v_3 \end{pmatrix} = \begin{pmatrix} a_{11}v_1 + a_{12}v_2 + a_{13}v_3 \\ a_{21}v_1 + a_{22}v_2 + a_{23}v_3 \\ a_{31}v_1 + a_{32}v_2 + a_{33}v_3 \end{pmatrix}. \tag{1.30}$$

Note that if two matrices multiply a vector, the matrix to the right operates on the vector first. That is, in $\mathcal{A}\,\mathcal{B}\,\vec{v}$, $\mathcal{B}$ operates first on $\vec{v}$, and then $\mathcal{A}$ operates on the vector result of $\mathcal{B}\,\vec{v}$. If $\mathcal{A}\,\mathcal{B}\,\mathcal{C}\,\vec{v}$, $\mathcal{C}$ operates before $\mathcal{B}$ and $\mathcal{A}$ operates after $\mathcal{B}$.

Transpose of a Matrix

The transpose of a matrix is produced by interchanging the subscripts of the matrix elements. That is, given

$$\mathcal{A} = \begin{pmatrix} a_{11} & a_{12} & a_{13} \\ a_{21} & a_{22} & a_{23} \\ a_{31} & a_{32} & a_{33} \end{pmatrix}, \tag{1.31}$$

the transpose of $\mathcal{A}$ is

$$\mathcal{A}^T = \begin{pmatrix} a_{11} & a_{21} & a_{31} \\ a_{12} & a_{22} & a_{32} \\ a_{13} & a_{23} & a_{33} \end{pmatrix}. \tag{1.32}$$

Inverse of a Matrix

The identity matrix is the matrix with 1's on the diagonal and all other elements 0.

$$\mathcal{I} = \begin{pmatrix} 1 & 0 & 0 \\ 0 & 1 & 0 \\ 0 & 0 & 1 \end{pmatrix}. \tag{1.33}$$

The inverse of matrix $\mathcal{A}$ is C, where $\mathcal{A}\,C = I$. The notation is

$$C = A^{-1}. \tag{1.34}$$

Length of a Vector

The length of a vector is the square root of the vector dotted with itself. That is, the length of vector v is

$$| \vec{v} | = \sqrt{\vec{v} \circ \vec{v}}. \tag{1.35}$$

A vector of length one is called a unit vector and usually has a "hat" over it instead of an arrow:

$$| \hat{v} | = \sqrt{\hat{v} \circ \hat{v}} = 1. \tag{1.36}$$

A Vector Field and Matrix Field

A vector field is a vector defined at each point in space. For example,

$$\vec{v} = \left(2x + 3y, \; 2x + y + z, \; x^2 + 3y - z^3\right) \tag{1.37}$$

is a vector field, because at each point (x, y, z) a vector defined.

A matrix field is the same idea as a vector field. For each point in the space, there is a different matrix. An example of a matrix field is

$$\mathscr{B} = \begin{pmatrix} 2y & x^2 + 4 & 7 \\ 6x - z & z^2 & 2x + 3 \\ 7y & 6z^2 & x^3 + y + z^2 \end{pmatrix}, \tag{1.38}$$

so that for each point (x, y, z) there is a matrix defined at that point.

A Matrix Times a Vector Field

The only difference between a matrix times a vector and a matrix times a vector field is that the components of the vector may be different at each point in space. Thus if

$$\mathscr{B} = \begin{pmatrix} 2 & 1 & 3 \\ 4 & 6 & 7 \\ 9 & 3 & 2 \end{pmatrix} \tag{1.39}$$

and

$$\vec{v} = \begin{pmatrix} 2x + y \\ 2z^2 \\ 3 + y \end{pmatrix} \tag{1.40}$$

then

$$\mathscr{B}\vec{v} = \begin{pmatrix} 9 + 4x + 5y + 2z^2 \\ 21 + 6x + 10y + 12z^2 \\ 6 + 18x + 11y + 6z^2 \end{pmatrix}. \tag{1.41}$$

A Matrix Field Times a Vector Field

As with a matrix times a vector field, a matrix field times a vector field is just a matrix times a vector, but both the matrix and the vector are functions of the point (x, y, z). Thus if

$$\mathscr{B} = \begin{pmatrix} 2x & y & 3(x+y) \\ 4y & 6z^2 & 7x \\ 9 & 3y & 2x \end{pmatrix} \tag{1.42}$$

and

$$\vec{v} = \begin{pmatrix} x \\ 2z^2 \\ 3 \end{pmatrix} \tag{1.43}$$

then

$$\mathscr{B}\vec{v} = \begin{pmatrix} 2x^2 + 9x + 9y + 2yz^2 \\ 21x + 4xy + 12z^4 \\ 15x + 6yz^2 \end{pmatrix}. \tag{1.44}$$

Calculus-Based Physics 1 Review

Newton's Laws in Equation Form

Newton's three laws can be reduced to two equations:

1. The sum of the forces on a body is equal to the mass times the acceleration of the body:

$$\Sigma \vec{F} = m\vec{a}. \tag{1.45}$$

2. If a exerts a force on b, then b exerts the same force on a, but in the opposite direction:

$$\vec{F}_{a \to b} = -\vec{F}_{b \to a}. \tag{1.46}$$

Force, Work, and Energy

If a force, $\vec{F}$, is exerted on a body and the body moves from point a to point b, the work done is defined as the integral of the force over the path taken between a and b:

$$\text{Work} = \int_a^b \vec{F} \circ d\vec{s}. \tag{1.47}$$

Since the dot product of two vectors is a scalar, work is independent of the coordinate system chosen.

A special case is where $\vec{F}$ is a conservative force. $\vec{F}$ is a conservative force if the path integral from a to b is independent of the path. In that case a potential energy can be defined at each point E_a and E_b such that the change in potential energy ΔE is

$$\Delta E = (E_b - E_a) = -\text{Work}. \tag{1.48}$$

The energy stored is actually stored in a field. In the case of gravity, the energy is stored in the gravitational field. In the case of a spring, the energy is stored in the electric field between the atoms of the spring. The work done by the force of gravity on a particle is $-\Delta E$. The work done on a spring by whatever is deforming it is $-\Delta E$ of the spring.

For example, if I pick up a box of mass 2 kg along a straight line from $\vec{a} = (0, 0, 1\text{m})$ to $\vec{b} = (0, 0, 4\text{m})$ in a gravitational field,

$$\vec{F}_g = -mg\hat{z} = -(2\text{kg})(9.8\text{m/s}^2)\hat{z} = -19.6\text{N}\hat{z}, \tag{1.49}$$

the work done is

$$\text{Work} = \int_a^b \vec{F}_g \circ d\vec{s} = \int_{(0,0,1m)}^{(0,0,4m)} (0,\,0,\,-19.6N) \circ d\vec{z} = -\int_{1m}^{4m} 19.6Ndz$$

$$= -19.6N(4m - 1,\, m) = -58.8J. \tag{1.50}$$

The gravitational field is a conservative field, so

$$-58.8J = -(E_b - E_a) \tag{1.51}$$

or

$$E_b = E_a + 58.5J, \tag{1.52}$$

and the box now has an additional 58.8 J of energy.

In physics you learned that for conservative forces the relationship between energy and force is

$$\vec{F} = -\vec{\nabla}E = -\left(\frac{\partial E}{\partial x},\, \frac{\partial E}{\partial y},\, \frac{\partial E}{\partial z}\right). \tag{1.53}$$

So if the gravitational energy field is

$$E = mgz, \tag{1.54}$$

then the force at any point z is

$$\vec{F}_g = -\vec{\nabla}E = -\vec{\nabla}(mgz) = -\left(\frac{\partial mgz}{\partial x},\, \frac{\partial mgz}{\partial y},\, \frac{\partial mgz}{\partial z}\right) = -(0,\,0,\,mg)$$

$$= (0,\,0,\,-mg), \tag{1.55}$$

and the force, $\vec{F}_g$, is a force in the downward direction of magnitude m g.

Miscellaneous

The Difference Between Point, Point Vector, and a General Vector

A point is defined by two (in 2D) or three (in 3D) ordered points. For example,

$$\vec{p} = (2,\, 3,\, 7) \tag{1.56}$$

is a point in 3D.

A point vector can be defined as an arrow from the origin to p and will then be

$$\vec{p} = (2, 3, 7) \tag{1.57}$$

A general vector, $\vec{v}$, can also be defined as

$$\vec{v} = (2, 3, 7) \tag{1.58}$$

In each case the notation is the same, (2, 3, 7), but there is a subtle difference between these three items. The point (2, 3, 7) is just one point in 3D space. The point vector is an arrow with its tail at (0, 0, 0). This is an unusual "vector" in that the point vector has a well-defined tail location. A general vector (2, 3, 7) can be represented as an arrow, but the tail of the arrow can be placed anywhere.

Material Properties

Materials will be considered to have continuous properties regardless of how small a sample is taken. This of course is not true of *any* material. All materials are made up of atoms which are discrete. However, the assumption of continuity is a "useful fiction" because it allows us to define derivatives of material properties. In numerical applications the material is divided into small regions which are assigned a single value to represent each material property within this region. This "works" because atoms are incredibly small. Even a cube of material 10^{-3} cm on a side still contains billions and billions of atoms.

Isotropic and Anisotropic Materials

An isotropic material is one that has the same properties in all directions. This means you can rotate the material without finding any difference in its response to a deformation. An anisotropic material is one that is not isotropic. An example would be a material made up of layers of different materials with different properties.

Homogeneous and Non-homogeneous Materials

A homogeneous material is one which has the same properties throughout the material. In our applications that means you can cut out any small piece of the material, and it will deform exactly as any other small piece of the material. A non-homogeneous material would have regions of one material and regions of a different material separated in space, so that one piece of the material would have different properties than another one.

Taylor Expansion and Physical Interpretation

It is useful to find how a function varies in the neighborhood of a point. For example, a plot of the function, $f(x) = x^2$, near $x = 0$ is flat, i.e., could be described by a horizontal line. However, near the point $x = 1$, a plot of the function has a nonzero slope. The variation of a function near a point is found mathematically with a Taylor expansion (or Taylor series).

Given a function $f(x)$, the Taylor expansion around the point $x = a$ is

$$f(x) = f(a) + f'(a)(x - a) + \frac{f''(x)}{2!}(x - a)^2 + \frac{f''(x)}{3!}(x - a)^3 + \dots$$

$$+ \frac{f^{(n)}}{n!}(x - a)^n + \dots, \tag{1.59}$$

where f' is the derivative of $f(x)$ evaluated at the point $x = a$, $f''(a)$ the second derivative evaluated at $x = a$, $f'''(a)$ the third derivative evaluated at $x = a$, and $f^{(n)}$ the nth derivative of $f(x)$ evaluated at a.

Although there are an infinite number of terms in the Taylor expansion, we will be interested only in points "near" $x = a$, where we can make the approximation,

$$f(x) \approx f(a) + f'(a)(x - a) \tag{1.60}$$

As an example, the function $f(x) = x^4$ evaluates near $x = 1$ as

$$f(x) = f(1) + f'(1)(x - 1) = 1^4 + 4(1)^3(x - 1) = 4x - 3. \tag{1.61}$$

So that a point "near" $x = 1$, say $x = 1.001$, has the approximate value from the Taylor's expansion to be 1.004. The actual value is 1.00401, a difference of only about 0.001%.

Derivative of a Vector by a Scalar

Here define the derivative of the vector $\vec{v}$ with respect to t as $\frac{\partial \vec{v}}{\partial t}$. The most common example in introductory physics is the acceleration, $\vec{a}$, in terms of the velocity, $\vec{v}$:

$$\vec{a} = \frac{\partial \vec{v}}{\partial t} = \left(\frac{\partial v_x}{\partial t}, \frac{\partial v_y}{\partial t}, \frac{\partial v_z}{\partial t} \right) \tag{1.62}$$

Note that t can be any scalar. In particular, although it may seem peculiar, the derivative of $\vec{v}$ with respect to $\mathcal{F}_{ij}$ could be defined as $\frac{\partial \vec{v}}{\partial \mathcal{F}_{ij}}$, where $\mathcal{F}_{ij}$ is an element of the $\mathcal{F}$ matrix, e.g.,

$$\vec{q} = \frac{\partial \vec{v}}{\partial \mathcal{F}_{12}} = \left(\frac{\partial v_x}{\partial \mathcal{F}_{12}}, \frac{\partial v_y}{\partial \mathcal{F}_{12}}, \frac{\partial v_z}{\partial \mathcal{F}_{12}} \right) \qquad (1.63)$$

Problems

Problem 1 Given
$$\vec{a} = (2, 3, 1)$$
$$\vec{b} = (3, 0, 2).$$
Find the following:

(a) $2\vec{a} + \vec{b}$

(b) $\vec{a} \circ \vec{b}$

(c) $\vec{a} \times \vec{b}$

(d) $\vec{a} \circ \left(\vec{b} \times \vec{c} \right)$

(e) $|\vec{a}|$.

Problem 2 Given $\vec{a}$ and $\vec{b}$ from problem 1 and
$$\mathcal{A} = \begin{pmatrix} 1 & .2 & .5 \\ -2.3 & 4 & 1.2 \\ 2.4 & .3 & .3 \end{pmatrix}$$
$$\mathcal{B} = \begin{pmatrix} 1 & 0 & .3 \\ 0 & 1 & .2 \\ -.3 & 2 & 1 \end{pmatrix}.$$
Find the following:

(a) $\mathcal{A} \, \mathcal{B}$

(b) $\mathcal{B} \, \mathcal{A}$

(c) $\mathcal{A} \, \vec{a} + \vec{b}$

(d) $\mathcal{A} \, \mathcal{B} \, \vec{b} - 2\vec{a}$.

Problem 3 Find the transpose of $\mathcal{A}$, where $\mathcal{A}$ is defined in problem 2.

Problem 4 Find the inverse of matrix $\mathcal{A}$, where $\mathcal{A}$ is defined in problem 2.

Problem 5 Given the vector field, $v = (2x + y, 3z - x, xy)$ and $\mathcal{A}$ from problem 2, find $\mathcal{A} v$ at the point (2,3,4).

Problem 6 A 36-kg boy pushes a 20 kg box with a force of 2 N. Both are on ice where there is no friction. What is the acceleration of both the boy and the box?

Problem 7 The boy in problem 6 picks up the box in problem 6 and places it on a table 2 m above the floor, 6 m from the original position of the box. How much work does the boy do?

Problem 8 A spring is defined to have a stored energy of $1/2\,k\chi^2$, where $k = 2$ N/m and χ is defined as the final length of the spring, L, minus the rest length of the spring, L_o. Assume one end of the spring is fixed at $x = 0$ and the other end is at rest at $x = 2$ m. If the moveable end of the spring is displaced to $x = 5$ m: (a) What is the change of energy in the spring? (b) How much work has been done by the spring? (c) What is the final force on the spring?

Problem 9 Find the Taylor expansion of $f(x) = 3(x - 5)(x^3 + 4)$ using Eq. 1.60 about the point $x = 2$. What is $f(2.01)$ in this approximation? What is the true value of f (2.01)?

Problem 10 Given $v = (2t + 4, t^2 4)$ and $t = x^3$, find (a) dv/dt and (b) dv/dx.

Chapter 2
Deformations

We will be interested in the deformation of a continuous body, e.g., the body shown in Fig. 2.1. We will begin with the body in some initial configuration in space and then let the body deform to some final configuration. Every point in the initial body will exist somewhere in the final body. We can relate the location of the initial points to the final points using a mapping. This chapter will describe a mapping and its application to deformations of materials.

A Map

A mapping provides a relationship between one set of points and another set of points. The mapping may be from one three-dimensional space to another three-dimensional space, or from a two-dimensional space to another two-dimensional space, or even from a one-dimensional space to another one-dimensional space. Of course, there are no one- or two-dimensional material objects, but one- and two-dimensional mappings are easier to see on a flat page and often carry enough information to illustrate concepts of maps in three-dimensional space. For any mapping that we are interested in, every point in the original space is mapped into a single new point in the mapped space, and every mapped point has been mapped from a single point in the original space. This is called a one-to-one mapping. For this type of mapping, the coordinates of each mapped point are functions of the coordinates of the corresponding initial point. Eventually these mappings will correspond to mappings from the original points occupied by a material that has not been deformed to the corresponding points in a material that has been deformed.

Supplementary Information The online version contains supplementary material available at [https://doi.org/10.1007/978-3-031-09157-5_2].

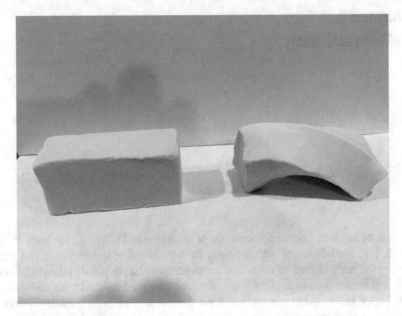

Fig. 2.1 Clay before and after deformation

Mathematics of a Mapping

Let us start with the simplest case of a one-dimensional map. The points in the original space can be described by the points on a number line, for example, all the points in line segment between 1 and 3. Each of these points in the original space can be mapped to a unique point in the new space. For example, each point between 1 and 3, $\{x \mid 1 < x < 3\}$, might be mapped onto the points between 2 and 6, $\{y \mid 2 < y < 6\}$, so that the mapping is $y = 2x$. With this mapping, the point at $x = 1$ is mapped to the point $y = 2$, the point at $x = 2$ is mapped onto the point $y = 4$, the point at $x = 2.5$ is mapped to $y = 5$, etc. To make notation more useful, a capital "x", X, will be used to represent any point in the original space and a small "x", x, to represent any point in the new space. Thus, the mapping in this example is $x = 2X$. Figure 2.2 displays this map.

The mapping in Fig. 2.2 is called an affine map, where an affine map is any map that can be described as

$$x = aX + b, \tag{2.1}$$

where a and b are constants. In Fig. 2.2, $a = 2$ and $b = 0$.

A non-affine map cannot be described with the form of Eq. 2.1. An example is $x = X^2$. Figure 2.3 illustrates an example of this map. Notice here that $X = 1$ maps to $x = 1$, $X = 2$ maps to $x = 4$, and $X = 3$ maps to $x = 9$, etc.

Fig. 2.2 A
one-dimensional map
defined as $x = 2X$

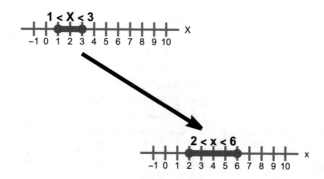

Fig. 2.3 A non-affine map,
defined as $x = X^2$

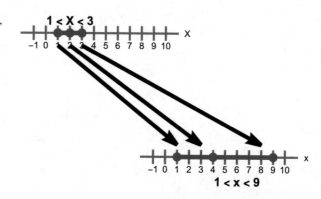

Now consider a two-dimensional map. For example, suppose the original space is $1 < X_1 < 3$ and $1 < X_2 < 3$, where any point in the space which is located by the vector $\vec{X} = (X_1, X_2)$ will after the mapping be located at $\vec{x} = (x_1, x_2)$, where $x_1 = 2X_1$ and $x_2 = 2X_2$. In this example, the point at $(2, 2)$ is mapped to the point $(4, 4)$. The point at $(1, 2)$ is mapped to the point $(2, 4)$, etc. Figure 2.4 illustrates this map. Notice here that the subscripts do not correspond to point numbers, but to coordinate axes – "1" corresponds to the abscissa, and "2" corresponds to the ordinate.

The relationship between the components of each point before and after the mapping could be more complicated. Examples are shown in Figs. 2.5, 2.6, and 2.7 along with the equations which define each mapping.

Notice that for the 2D mapping, the coordinates of each point can be functions of both X_1 and X_2. In Fig. 2.4 the new mapping is drawn in a displaced coordinate system so that the initial and final points do not overlap. In Fig. 2.5 the initial and final maps are displaced and can be displayed on a single X_1, X_2 coordinate system without overlap. For a mapping of real material, the coordinate axes will always be the same coordinates axes before and after the mapping. This set of coordinate axes define a fixed coordinate system.

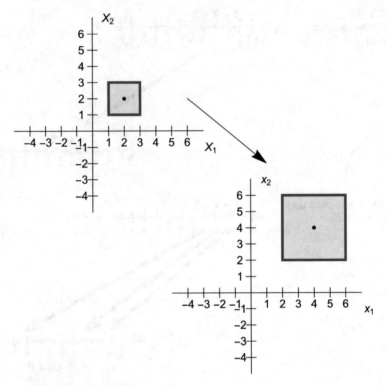

Fig. 2.4 A two-dimensional mapping of $1 \leq X_1 \leq 3$ and $1 \leq X_2 \leq 3$ defined by $x_1 = 2X_1$ and $x_2 = 2X_2$

Figure 2.5 is an affine map in 2D since it is of the form

$$x_1 = m_{11}X_1 + m_{12}X_2 + b_1$$
$$x_2 = m_{21}X_1 + m_{22}X_2 + b_2, \tag{2.2}$$

where m_{11}, m_{12}, m_{21}, m_{22}, b_1, and, b_2 are all constants.

Non-affine maps can also be illustrated in 2D as shown in Fig. 2.6. Notice that the mapping in Fig. 2.6 cannot be described using Eq. 2.2. In general, all that can be said about a non-affine mapping is that the coordinates of any final point, (x_1, x_2), are functions of a corresponding point, (X_1, X_2). This can be written as

$$x_1 = f_1(X_1, X_2)$$
$$x_2 = f_2(X_1, X_2), \tag{2.3}$$

which can be shortened further by writing Eq. 2.3 as

$$f_i(X_1, X_2) = x_i(X_1, X_2) \quad \text{for } i = 1 \text{ and } 2. \tag{2.4}$$

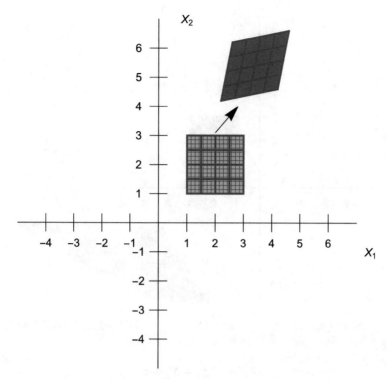

Fig. 2.5 The mapping from $1 \leq X_1 \leq 3$ and $1 \leq X_2 \leq 3$ defined by $x_1 = f_1(X_1, X_2) = X_1 + 0.2 X_2 + 1$ and $x_2 = f_2(X_1, X_2) = 0.2X_1 + X_2 + 3$. Here the displacement of the space is large enough that the initial and final set of points do not overlap and can therefore be shown on the same coordinate system

Maps can also be piecewise. The map illustrated in Fig. 2.7 maps $X_1 < 2$ with a different function than those points with $X_1 \geq 2$. In Fig. 2.7, the mapping is as follows:

$$\text{If}(X_1 < 2) \quad \text{then} \quad x_1 = X_1 X_2 + .2X_2 \quad \text{and} \quad x_2 = X_1^{1.3} + 01X_2$$
$$\text{If}(X_1 \geq 2) \quad \text{then} \quad x_1 = 2X_1 + X_2 \quad \text{and} \quad x_2 = X_1^{1.3} + 0.1X_2 + 1. \tag{2.5}$$

Piecewise mappings can be used to describe materials that are both deformed and broken into more than one piece.

All of these forms can be repeated in 3D, but they are harder to illustrate on a two-dimensional page. As a result, I will show only one 3D affine map in Fig. 2.8 and one 3D non-affine map in Fig. 2.9. The affine map in Fig. 2.8 is defined as follows:

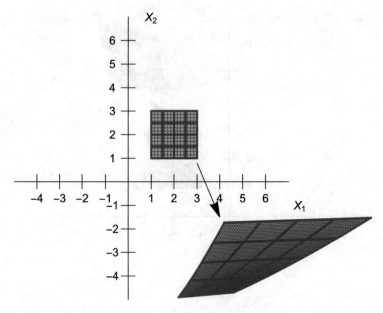

Fig. 2.6 The mapping from $1 \leq X_1 \leq 3$ and $1 \leq X_2 \leq 3$ defined by $x_1 = f_1(X_1, X_2) = X_1 X_2 + 0.2X_2 + 1$ and $x_2 = f_2(X_1, X_2) = X_1^{1.3} + 0.1X_2 - 6$

$$
\begin{aligned}
x1 &= 2X_1 + X_2 + 4 \\
x2 &= 2X_1 + 0.1X_2 + 1 \\
x3 &= X_1 + 0.5X_3,
\end{aligned} \tag{2.6}
$$

where a general 3D affine map is always of the form

$$
\begin{aligned}
x_1 &= m_{11}X_1 + m_{12}X_2 + m_{13}X_3 + b_1 \\
x_2 &= m_{21}X_1 + m_{22}X_2 + m_{23}X_3 + b_2 \\
x_3 &= m_{31}X_1 + m_{32}X_2 + m_{33}X_3 + b_3
\end{aligned} \tag{2.7}
$$

and here

$$
\begin{aligned}
&m_{11} = 2,\ m_{12} = 1,\ m_{13} = 0,\ b_1 = 4 \\
&m_{21} = 2,\ m_{22} = 0.1,\ m_{23} = 0,\ b_2 = 1 \\
&m_{31} = 1,\ m_{32} = 0,\ m_{33} = 0.5,\ b_3 = 0.
\end{aligned} \tag{2.8}
$$

The non-affine map in Fig. 2.9 is defined as follows:

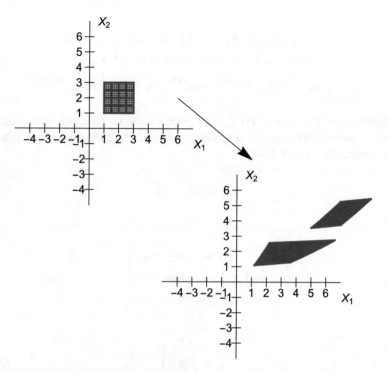

Fig. 2.7 A piecewise affine mapping in 2D

Fig. 2.8 An affine mapping in 3D

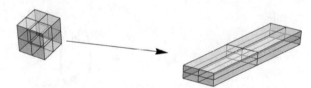

Fig. 2.9 A 3D non-affine mapping

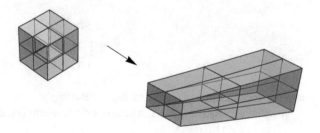

$$x_1 = f_1(X_1, X_2, X_3) = x_1(X_1, X_2, X_3) = 2X_2 + X_1$$
$$x_2 = f_2(X_1, X_2, X_3) = x_2(X_1, X_2, X_3) = X_2^{1.3} + 0.1X_1 + 1 \qquad (2.9)$$
$$x_3 = f_3(X_1, X_2, X_3) = x_3(X_1, X_2, X_3) = \frac{1}{2}X_2X_3.$$

Equations of the form of Eq. 2.7 are cumbersome to write out, so a couple of notations are used to simplify the writing of these. One form is the matrix notation form, where Eq. 2.7 is written as

$$\vec{x} = \mathcal{M}\vec{X} + \vec{b} \qquad (2.10)$$

with

$$\mathcal{M} = \begin{pmatrix} m_{11} & m_{12} & m_{13} \\ m_{21} & m_{22} & m_{23} \\ m_{31} & m_{32} & m_{33} \end{pmatrix} \qquad (2.11)$$

$$\vec{X} = \begin{pmatrix} X_1 \\ X_2 \\ X_3 \end{pmatrix} \qquad (2.12)$$

$$\vec{b} = \begin{pmatrix} b_1 \\ b_2 \\ b_3 \end{pmatrix} \qquad (2.13)$$

$$\vec{x} = \begin{pmatrix} x_1 \\ x_2 \\ x_3 \end{pmatrix} \qquad (2.14)$$

and the normal rules of matrix multiplication apply.

Another alternative is the Einstein notation, where Eq. 2.7 is written as

$$x_i = m_{ij}X_j + b_i, \qquad (2.15)$$

where i and j take on values 1, 2, and 3, and it is understood that repeated indices in products are summed over. Thus Eqs. 2.7, 2.10, and 2.15 all are equivalent and can be used interchangeably to represent an affine 3D map.

The Application of a Mapping

Most material deformations are not affine, so you might wonder what is the utility of the affine map in a theory of general deformations of a material body. The answer is that even non-affine maps are locally affine. To see what is meant by this, consider the non-affine mapping shown in Fig. 2.3,

$$x = f(X) = X^2. \qquad (2.16)$$

This mapping cannot be represented by Eq. 2.15, but a Taylor expansion of X^2 about any point X_0 yields

$$x = f(X_0) + \frac{df}{dx}\Big|_{X=X_0} (X - X_0) + O\left((X - X_0)^2\right). \qquad (2.17)$$

which for first order in $(X - X_0)$,

$$x \approx f(X_0) + \frac{df}{dX}\Big|_{X=X_0} (X - X_0) \qquad (2.18)$$

For $x = X^2$, the result is as follows:

$$x \approx X_0^2 + 2X_0(X - X_0) = (2X_0)X - X_0^2. \qquad (2.19)$$

At each point, X_0, the mapping near X_0 is affine with $M = +2X_0$ and $b = -X_0^2$. This result is only correct, "near" the point X_0. This concept can be captured better by taking the derivative of both sides of Eq. 2.16. Then

$$dx = \frac{df}{dX} dX, \qquad (2.20)$$

which is an affine mapping between dX and dx. Note that $\frac{df}{dX}$ is different for each point, X, but is a constant for any particular X. This is what is meant by the mapping being locally affine. This is easily extended to 2D and 3D.

Derivatives of $x_i(X_1, X_2)$ and $x_i(X_1, X_2, X_3)$ yield locally affine maps for two- and three-dimensional maps. In particular for 2D

$$dx_i = \frac{\partial x_i}{\partial X_1} dX_1 + \frac{\partial x_i}{\partial X_2} dX_2 \text{ for } i = 1 \text{ and } 2, \qquad (2.21)$$

where

$X_1 =$ the x component of the point the derivatives are defined
$X_2 =$ the y component of the point the derivatives are defined.

or using the Einstein summation notation,

$$dx_i = \frac{\partial f_i}{\partial X_j} dX_j \quad \text{for } i = 1 \text{ and } 2 \text{ and } j \text{ summed over } 1 \text{ and } 2. \quad (2.22)$$

A similar derivation in 3D yields

$$dx_i = \frac{\partial f_i}{\partial x_j} dX_j \quad \text{for } i = 1, 2, \text{ and } 3 \text{ and } j \text{ summed over } 1 \text{ through } 3. \quad (2.23)$$

Equation 2.23 can also be written

$$dX_i = \frac{\partial x_i}{\partial X_j} dX_j \quad \text{for } i = 1, 2, \text{ and } 3 \text{ and } j \text{ summed over } 1 \text{ through } 3. \quad (2.24)$$

The matrix $\frac{\partial x_i}{\partial X_j}$ is very useful in the description of deformations. It is called the deformation gradient matrix and is denoted as $\mathcal{F}$,

$$\mathcal{F}_{ij} = \frac{\partial x_i}{\partial X_j} \quad (2.25)$$

The local affine connection can be illustrated in 2D and 3D cases by drawing coordinate vectors in each direction and observe how they map into the new space. This is done by posting the origin of the coordinate vectors at the initial point and the mapped vectors at the corresponding mapped point.

As an example, consider the nonlinear map

$$\begin{aligned} x_1 &= -X_1 X_2 \sin{(0.8 - X_1)} + 0.2X_2 + 1 \\ x_2 &= -2 - X_1 - 0.00625(X_2 - 5)^4 \end{aligned} \quad (2.26)$$

shown in Fig. 2.10. The partial derivatives of Eq. 2.26 give

$$\frac{\partial x_1}{\partial X_1} = X_1 X_2 \cos{(0.8 - X_1)} - X_2 \sin{(0.8 - X_1)} \quad (2.27)$$

$$\frac{\partial x_1}{\partial X_2} = +0.2 - -X_1 \sin{(0.8 - X_1)}, \quad (2.28)$$

$$\frac{\partial x_2}{\partial X_1} = -1.3X_1^{0.3}, \quad (2.29)$$

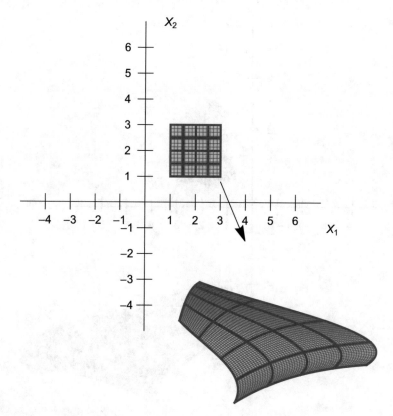

Fig. 2.10 A non-affine map in 2D

$$\frac{\partial x_2}{\partial X_2} = -0.025(X_2 - 5)^3, \tag{2.30}$$

When these are substituted back into Eq. 2.22 with $X_1 = 1$ and $X_2 = 1$, we get

$$
\begin{aligned}
dx_1 &= 1.17874 \, dX_1 + 0.398669 \, dX_2 \\
dx_2 &= -1.3 \, dX_1 + 1.6 \, dX_2.
\end{aligned} \tag{2.31}
$$

The vectors that lie along the coordinate axes are posted in Fig. 2.11, i.e., $\vec{dX}^{(1)} = (dX_1, 0)$ and $\vec{dX}^{(2)} = (0, dX_2)$. These vectors are posted at the point the partial derivatives are defined, i.e., $X_0 = (1, 1)$. Using Eq. 2.22, the vector $\vec{dX}^{(1)}$ maps into the vector

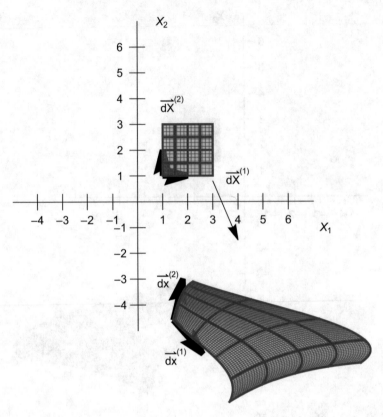

Fig. 2.11 The dX vectors lying along the coordinate axes are illustrated at $(1, 1)$. These vectors mapped into the new region are also shown

$$\overset{\rightarrow (1)}{dX} = (dx_1, dx_2)$$
$$= \left(\frac{\partial x_1}{\partial x_1} dX_1 + \frac{\partial x_1}{\partial X_2} dX_2, \frac{\partial x_2}{\partial X_1} dX_1 + \frac{\partial x_2}{\partial x_2} dX_2 \right). \tag{2.32}$$

For $dX_2 = 0$,

$$\overset{\rightarrow (1)}{dX} = (dx_1, dx_2)$$
$$= \left(\frac{\partial x_1}{\partial X_1}, \frac{\partial x_2}{\partial X_1} \right) dX_1 \tag{2.33}$$
$$= (1.17874, -1.3) dX_1.$$

Similarly vector $dX^{(2)}$ maps into the vector

$$\overrightarrow{dX}^{(2)} = (dx_1, dx_2)$$
$$= \left(\frac{\partial x_1}{\partial X_1} dX_1 + \frac{\partial x_1}{\partial X_2} dX_2, \frac{\partial x_2}{\partial X_1} dX_1 + \frac{\partial x_2}{\partial X_2} dX_2\right). \tag{2.34}$$

For $dX_1 = 0$,

$$\overrightarrow{dX}^{(2)} = (dx_1, dx_2)$$
$$= \left(\frac{\partial x_1}{\partial X_2}, \frac{\partial x_2}{\partial X_2}\right) dX_2 \tag{2.35}$$
$$= (0.398669, 1.6) dX_2.$$

By choosing values of $dX_1 = 1$ and $dX_2 = 1$, these vectors are illustrated in Fig. 2.11.

Figure 2.12 repeats the process used to create Fig. 2.11 for the point (2, 2). Notice that both the lengths and directions of $dx^{(1)}$ and $dx^{(2)}$ change as we move from point to point in the mapping. But locally (i.e., near each chosen point), the mapping is

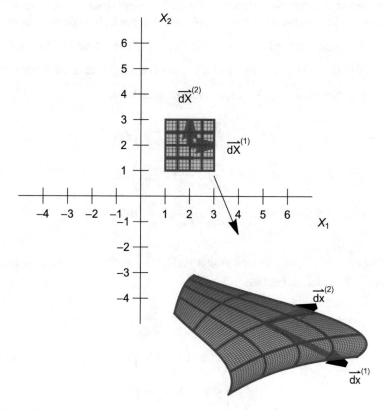

Fig. 2.12 The dX vectors lying along the coordinate axes are illustrated at (2, 2). These vectors mapped into the new region are also shown

Fig. 2.13 Posting of $dX^{(i)}$ and $dx^{(i)}$ for $i = 1, 2,$ and 3 for the mapping shown

$$x_1 = 2X_2 + X_1$$

$$x_2 = X_2^{1.3} + 0.1\,X_1 + 1$$

$$x_3 = \frac{1}{2}\,X_2 X_3$$

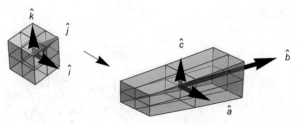

affine. That is, the affine equations (Eq. 2.22) describe the mapping of the points "near" (1, 1), and a different linear map describes the mapping of the points "near" (2, 2).

Figure 2.13 shows the result of repeating the process of displaying coordinate vectors before and after deformation in 3D for the mapping shown there. In 3D, the three vectors that lie along the three coordinate axes can be thought of as bounding either a small cuboid or a small tetrahedron defined by $\overset{\rightarrow (1)}{dX} = (dX_1, 0, 0)$, $\overset{\rightarrow (2)}{dX} = (0, dX_2, 0)$, and $\overset{\rightarrow (3)}{dX} = (0, 0, dX_3)$). Each of these vectors are mapped into the $\overset{\rightarrow (1)}{dX}, \overset{\rightarrow (2)}{dX}, \overset{\rightarrow (3)}{dX}$ vectors as follows:

$$\overset{\rightarrow (1)}{dX} = \left(\frac{\partial x_1}{\partial X_1}, \frac{\partial x_2}{\partial x_1}, \frac{\partial x_3}{\partial x_1}\right) dX_1$$

$$\overset{\rightarrow (2)}{dX} = \left(\frac{\partial x_1}{\partial X_2}, \frac{\partial x_2}{\partial X_2}, \frac{\partial x_3}{\partial X_2}\right) dX_2 \qquad (2.36)$$

$$\overset{\rightarrow (3)}{dX} = \left(\frac{\partial x_1}{\partial x_3}, \frac{\partial x_2}{\partial x_3}, \frac{\partial x_3}{\partial x_3}\right) dX_3.$$

If the $\overset{\rightarrow (1)}{dX}$ vectors are chosen as unit vectors, $\hat{\imath}, \hat{\jmath}, \hat{k}$, these vectors map into the vectors $\vec{a}, \vec{b},$ and $\vec{c}$ as follows:

$$\vec{a} = \left(\frac{\partial x_1}{\partial X_1}, \frac{\partial x_2}{\partial X_1}, \frac{\partial x_3}{\partial X_1}\right)$$

$$\vec{b} = \left(\frac{\partial x_1}{\partial x_2}, \frac{\partial x_2}{\partial x_2}, \frac{\partial x_3}{\partial x_2}\right) \qquad (2.37)$$

$$\vec{c} = \left(\frac{\partial x_1}{\partial x_3}, \frac{\partial x_2}{\partial x_3}, \frac{\partial x_3}{\partial x_3}\right).$$

An example of this for the mapping used in Fig. 2.9 is shown in Fig. 2.13. Note that the $\vec{a}$, $\vec{b}$, and $\vec{c}$ vectors are the column vectors of $\mathcal{F}$ defined in Eq. 2.25.

A mathematical description of the deformation of each point, a mapping, and a mathematical description of how the body deforms near each point are now defined. The deformation near each point will be used to calculate the internal forces within each region. This, along with the original mapping, provides the equations for the deformation of an elastic material body. Before we complete this development, however, we must first describe the forces acting on the body in Chap. 3.

Problems

Problem 1 Plot the mapping $x = 2X^3 + 3X + 2$ with $1 \leq X \leq 4$ as in Fig. 2.2.

Problem 2 Which of the following 1D maps are affine?

(a) $x = 2X + 3 + 7X$
(b) $x = X^2 + 3X$
(c) $x = 2$
(d) $x = 3X..$

Problem 3 Find where the point (2.5, 3) is mapped into by using the equations given for the mapping in Fig. 2.4.

Problem 4 Rewrite the mapping in Fig. 2.5 using the form of Eq. 2.2, and explain why this mapping is not an affine map.

Problem 5 Find a 2D affine map that takes the points (1, 1), (2, 3), and (4, 1) into the points (2, 2), (6, 1), and (8, 3), respectively.

Problem 6 Given the map below, what point does the point (4, 3) map into?

$$x_1 = 2X_1 + 3X_2 - 3$$
$$x_2 = 5X_2.$$

Problem 7 Using the following map, find $\frac{\partial x_1}{\partial x_1}$, $\frac{\partial x_1}{\partial x_2}$, $\frac{\partial x_2}{\partial x_1}$, $\frac{\partial x_2}{\partial x_2}$

$$x_1 = X_1^3 - 2X_2 + 3$$
$$x_2 = X_1 X_2.$$

Problem 8 Given the map in problem 7, find the vectors that are mapped from $d\vec{X}^{(1)} = \{1, 0\}$ and $d\vec{X}^{(2)} = \{0, 1\}$.

Chapter 3
Forces

This chapter describes the forces acting within a deformed body. The goal here is to produce an equation of motion of each region within the deformed body in terms of the local forces. To accomplish this, first divide the body into small cubes that align with the coordinate axes. That is, for each initial cube, there will be a "top" surface oriented so that the normal vector points in the +z direction, a "bottom" surface with a normal vector pointing in the $-z$ direction, and side surfaces that have normal vectors pointing in the $+x$, $-x$, $+y$, and $-y$ directions. These cubes will deform as the body deforms. Figure 3.1 shows an example of this before and after a deformation. The forces acting on the cubes will change with the deformation and may be functions of time as well as space. The sum of the forces acting on each region will result in the mass times acceleration of each small region of space in a deforming body.

We will divide the forces on each cube into two categories – body forces and surface forces. Body forces are forces that act throughout the material. The most common example is gravity. Surface forces act on the surface of the body or on the surface of any sub-piece of the body. Figure 3.2 shows a small cubical region within the material. Surface forces may act on all six surfaces of the cube. Shown in Fig. 3.2 are only the forces acting on the front surfaces of a cube, along with a body force shown acting at the center of the cube. We will concentrate on the surface forces first. Later we will add back the body forces before we end the chapter.

Consider first the surface force on the top of the cube. This force can be divided into x, y, and z components as shown in Fig. 3.3.

The force on each face of the cube can also be divided into its x, y, and z components. The following notation is used to describe the components of the forces on the surfaces normal to +x, +y, and +z:

For the +x surface of the cube:

Supplementary Information The online version contains supplementary material available at [https://doi.org/10.1007/978-3-031-09157-5_3].

Fig. 3.1 A material region
divided into cubes. These
cubes before and after an
example deformation are
shown

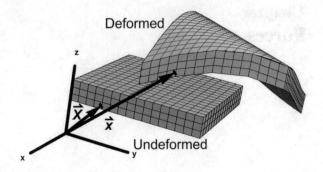

Fig. 3.2 Surface forces on
the front faces and a body
force on the center of a
single cube within a body

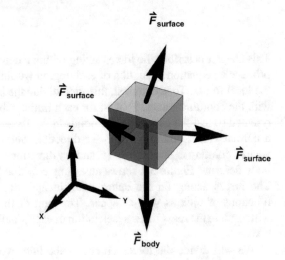

Fig. 3.3 Dividing a surface
force into its components

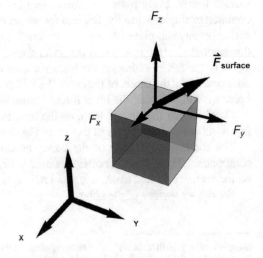

Fig. 3.4 The components of the forces on each of the surfaces with normals to the positive coordinate axes

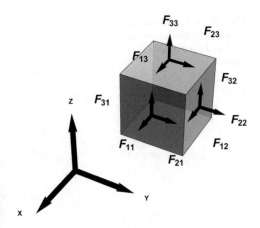

F_{11} is the x component of the force acting on the front of the cube.
F_{21} is the y component of the force acting on the front of the cube.
F_{31} is the z component of the force acting on the front of the cube.

For the +y surface of the cube:

F_{12} is the x component of the force acting on the side of the cube.
F_{22} is the y component of the force acting on the side of the cube.
F_{32} is the z component of the force acting on the side of the cube.

For the +z surface of the cube:

F_{13} is the x component of the force acting on the top of the cube.
F_{23} is the y component of the force acting on the top of the cube.
F_{33} is the z component of the force acting on the top of the cube.

Notice that the first subscript on F represents the direction the force is aligned with 1 being the *x* direction, 2 being the y direction, and 3 being the *z* direction. The second subscript identifies the surface the force is applied to: 1 being the surface initially parallel to the y-z surface, i.e., the surface represented by a vector parallel to the x coordinate axis, 2 being the surface with surface vector parallel to the y coordinate axis, and 3 being the surface with surface vector parallel to the z coordinate axis. Figure 3.4 illustrates these notations.

As the material deforms, the initial surfaces of the cube may no longer align with the coordinate axes. When that happens, forces can still be divided into components in the x, y, and *z* directions for each of the cube's surfaces. Figure 3.5 shows an example of this on a deformed top surface.

Now consider only the y component of the force against the y plane as shown in Fig. 3.6. In Fig. 3.6 the body has been extended in the +y direction. To shorten the notation, define the following:

Fig. 3.5 Dividing a surface force into its components on the initial top surface of a cube

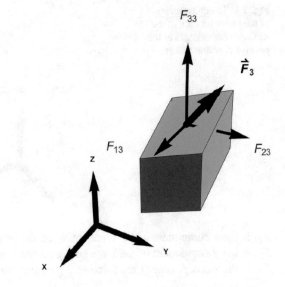

Fig. 3.6 The F_{22} forces at (X_1, X_2, X_3, t) and $(X_1, X_2 + X_2, X_3, t)$

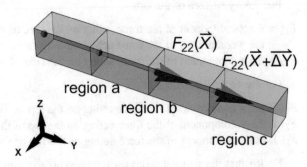

$$
\begin{aligned}
F_{ij}\left(\vec{X}\right) &= F_{ij}(X_1, X_2, X_3, t) \\
F_{ij}\left(\vec{X} + \vec{\Delta X}\right) &= F_{ij}(X_1 + \Delta X_1, X_2, X_3, t) \\
F_{ij}\left(\vec{X} + \vec{\Delta Y}\right) &= F_{ij}(X_1, X_2 + \Delta X_2, X_3, t) \\
F_{ij}\left(\vec{X} + \vec{\Delta Z}\right) &= F_{ij}(X_1, X_2, X_3 + \Delta X_3, t).
\end{aligned}
\tag{3.1}
$$

We will concentrate first on the forces acting in the y direction on the different faces of a cube of material. Using Eq. 3.1, $F_{22}\left(\vec{X}\right)$ is the force that region b exerts on region a. Not shown is the force - $F_{22}\left(\vec{X}\right)$ that region a exerts on region b. (Remember Newton's third law: If "a" exerts a force on "b", then "b" exerts an equal and opposite force on "a".) Also shown is $F_{22}\left(\vec{X} + \vec{\Delta Y}\right)$, which is the force that region c exerts on region b. (Not shown is the force region b exerts on region c, $-F_{22}\left(\vec{X} + \vec{\Delta Y}\right)$.)

Fig. 3.7 Forces on region b
in the y direction on the y
faces of region b

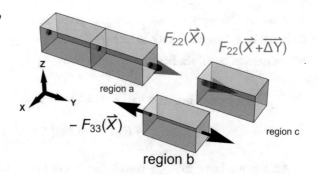

Fig. 3.8 Forces acting on
the top and bottom surfaces
of region b

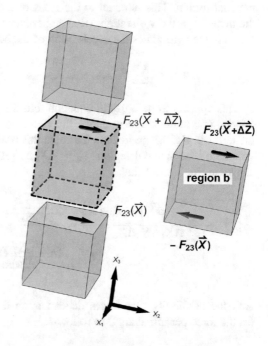

If region b is extracted in Fig. 3.6, the forces acting on this material are shown in Fig. 3.7, $F_{22}\left(\vec{X} + \overrightarrow{\Delta Y}\right)$ on the front face and $-F_{22}\left(\vec{X}\right)$ on the back face of region b. The total force on these two faces in the y direction is the sum of these forces, $F_{22}\left(\vec{X} + \overrightarrow{\Delta Y}\right) - F_{22}\left(\vec{X}\right)$.

Now let us concentrate on the top and bottom ($\pm z$ surfaces) surfaces of region b, Fig. 3.8. The top force in the y direction is $F_{23}\left(\vec{X} + \overrightarrow{\Delta Z}\right)$. The force on the bottom surface of region b is $-F_{23}\left(\vec{X}\right)$. The total force in the y direction on the top and bottom of region b is $F_{23}(\vec{X} + \overrightarrow{\Delta Z}) - F_{23}(\vec{X})$.

A similar description in the x direction gives the forces in the y direction on the front and back face of region b to be $F_{21}\left(\vec{X} + \vec{\Delta X}\right) - F_{21}\left(\vec{X}\right)$. Adding the forces in the y direction on all six faces of the cube gives

$$\sum F_{\text{surfaces in the y direction}} = F_{21}\left(\vec{X} + \vec{\Delta X}\right) - F_{21}\left(\vec{X}\right) + F_{22}\left(\vec{X} + \vec{\Delta Y}\right)$$

$$- F_{22}\left(\vec{X}\right) + F_{23}\left(\vec{X} + \vec{\Delta Z}\right) - F_{23}\left(\vec{X}\right). \tag{3.2}$$

Adding the body forces gives all the forces acting in the y direction on the material region. This is equal to the mass of the region times the acceleration of the material in the y direction. (Remember Newton's first and second law; sum of forces is the mass times the acceleration of the body.)

$$m a_y = \sum F_{\text{surfaces in the y direction}} + \sum F_{\text{body in the y direction}}. \tag{3.3}$$

This appears to be a useful equation, but the problem is that as the material is divided into more and more cubes, the volume of each cube shrinks to zero, and all the terms in Eq. 3.3 go to zero. To solve this problem, divide Eq. 3.3 by the initial volume of the material region, $V_0 = \Delta X_1 \Delta X_2 \Delta X_3$. For just the first two y terms shown in Eq. 3.2, we find

$$\frac{F_{21}\left(\vec{X} + \vec{\Delta X}\right) - F_{21}\left(\vec{X}\right)}{\Delta X_1 \Delta X_2 \Delta X_3} = \frac{\frac{F_{21}\left(\vec{X}+\vec{\Delta X}\right)}{\Delta X_1 \Delta X_2} - \frac{F_{21}\left(\vec{X}\right)}{\Delta X_2 \Delta X_3}}{\Delta X_1}$$

$$= \frac{\frac{F_{21}(X_1+\Delta X_1, X_2, X_3, t)}{\Delta X_2 \Delta X_3} - \frac{F_{21}(X_1, X_2, X_3, t)}{\Delta X_2 \Delta X_3}}{\Delta X_1} \tag{3.4}$$

Add the possibility that forces depend upon time and define engineering stress (as the force per initial area) as follows:

$$\mathcal{P}_{21}(X_1, X_2, X_3, t) = \frac{F_{21}(X_1, X_2, X_3, t)}{\Delta X_2 \Delta X_3} \tag{3.5}$$

and Eq. 3.4 is

$$\frac{F_{21}(X_1 + \Delta X_1, X_2, X_3, t) - F_{21}(X_1, X_2, X_3, t)}{\Delta X_1 \Delta X_2 \Delta X_3}$$

$$= \frac{\mathcal{P}_{21}(X_1 + \Delta X_1, X_2, X_3, t) - \mathcal{P}_{21}(X_1, X_2, X_3, t)}{\Delta X_1} \tag{3.6}$$

Repeating this for each surface, gives

$$\frac{1}{V_0} \sum F_{\text{surfaces in the y direction}} = \frac{\mathcal{P}_{21}(X_1 + \Delta X_1, X_2, X_3, t) - \mathcal{P}_{21}(x, y, z, t)}{\Delta X_1}$$
$$+ \frac{\mathcal{P}_{22}(X_1, X_2 + \Delta X_2, X_3, t) - \mathcal{P}_{22}(X_1, X_2, X_3, t)}{\Delta X_2} \qquad (3.7)$$
$$+ \frac{\mathcal{P}_{23}(X_1, X_2, X_3 + \Delta X_3, t) - \mathcal{P}_{23}(X_1, X_2, X_3, t)}{\Delta X_3}.$$

Taking the limit as V_0 approaches 0 (with ΔX_1, ΔX_2, and ΔX_3 also approaching zero) in Eq. 3.7, we get

$$\lim{}_{V_0 \to 0} \frac{1}{V_0} \sum F_{\text{surfaces in the y direction}} = \frac{\partial \mathcal{P}_{21(X_1, X_2, X_3, t)}}{\partial X_1} + \frac{\partial \mathcal{P}_{22(X_1, X_2, X_3, t)}}{\partial X_2}$$
$$+ \frac{\partial \mathcal{P}_{23(X_1, X_2, X_3, t)}}{\partial X_3}. \qquad (3.8)$$

In Einstein notation, Eq. 3.8 is

$$\lim{}_{V_0 \to 0} \frac{1}{V_0} \sum F_{\text{surfaces in the y direction}} = \frac{\partial \mathcal{P}_{2i}(X_1, X_2, X_3, t)}{\partial X_i}. \qquad (3.9)$$

Returning to Eq. 3.3, divide both sides by V_0, and let this volume approach zero to give

$$\lim{}_{V_0 \to 0} \frac{m}{V_0} a_y = \lim{}_{V_0 \to 0} \frac{1}{V_0} \sum F_{\text{surfaces in the y direction}}$$
$$+ \lim{}_{V_0 \to 0} \frac{1}{V_0} \sum F_{\text{surfaces in the y direction}}. \qquad (3.10)$$

If the only body force is gravity in the $-z$ direction, then

$$\sum F_{\text{body in the y direction}} = 0 \qquad (3.11)$$

$$\sum F_{\text{body in the x direction}} = 0 \qquad (3.12)$$

$$\sum F_{\text{body in the z direction}} = -mg. \qquad (3.13)$$

Define a mass density as

$$\rho_0 = \frac{m}{V_0}. \qquad (3.14)$$

Notice that this density, ρ_0, is not the usual definition of density, ρ. Normally density is mass divided by current volume, not mass divided by the initial volume.

For the y direction,

$$\rho_0 a_y = \frac{\partial \mathcal{P}_{2i}(X_1, X_2, X_3, t)}{\partial X_i} \quad \text{with } i \text{ summed from 1 to 3.} \quad (3.15)$$

Repeating this derivation for all three directions, adding the force of gravity in the $-z$ direction, and using the Einstein notation, we get

$$\rho_0 a_j = \frac{\partial \mathcal{P}_{ji(X_1,X_2,X_3,t)}}{\partial X_i} - \rho_0 g \delta_{j3}, \quad (3.16)$$

where

$$\delta_{ij} = 1 \quad \text{if } i = j \text{ and } 0 \; i \neq j \quad (3.17)$$

For clarity, Eq. 3.16 is written out in detail as follows:

$$\rho_0 a_x = \frac{\partial \mathcal{P}_{11(X_1,X_2,X_3,t)}}{\partial X_1} + \frac{\partial \mathcal{P}_{12(X_1,X_2,X_3,t)}}{\partial X_2} + \frac{\partial \mathcal{P}_{13(X_1,X_2,X_3,t)}}{\partial X_3} \quad (3.18)$$

$$\rho_0 a_y = \frac{\partial \mathcal{P}_{21(X_1,X_2,X_3,t)}}{\partial X_1} + \frac{\partial \mathcal{P}_{22(X_1,X_2,X_3,t)}}{\partial X_2} + \frac{\partial \mathcal{P}_{23(X_1,X_2,X_3,t)}}{\partial X_3} \quad (3.19)$$

$$\rho_0 a_z = \frac{\partial \mathcal{P}_{31(X_1,X_2,X_3,t)}}{\partial X_1} + \frac{\partial \mathcal{P}_{32(X_1,X_2,X_3,t)}}{\partial X_2} + \frac{\partial \mathcal{P}_{33(X_1,X_2,X_3,t)}}{\partial X_3} - \rho_0 g. \quad (3.20)$$

Equations 3.18, 3.19, and 3.20 are the equations of motion of a deforming material. What is next is to relate the surface forces to the deformations.

Problems

Problem 1 An incompressible cube of size 2m on each side is positioned so that the bottom surface of the cube lies on the x-y plane. The corner of the cube is at $(0, 0, 0)$, and the planes of the cube are parallel to the coordinate planes. A total force of 2 N is applied downward in the $-z$ direction to the top surface of the cube. Find $\mathcal{P}_{13}$, $\mathcal{P}_{23}$, , and $\mathcal{P}_{33}$ on the top of the cube.

Problem 2 The cube in problem 1 is replaced by a non-homogeneous, compressible cube. As a result of the applied force in the $-z$ direction, the top surface of the cube is now defined by the four points (0, 0, 2m), (2m, 0, 2m), (2m, 2m, 1m), and (0, 2m, 1m). Assuming the applied force has not changed, find $\mathcal{P}_{13}$, $\mathcal{P}_{23}$, and $\mathcal{P}_{33}$ on the top of the cube now. (Hint: The initial top surface is the same as that in problem 1.)

Problem 3 If the cube in problem 1 is sitting stationary with no gravity and the only applied force is 2 N downward, the force per unit original area must be uniform throughout the cube. (Otherwise, $\partial f_{33}/\partial X_3 \neq 0$ and the cube would accelerate.) Given this, how is it that the force on the top of the cube is *downward*, but the force on the bottom of the cube from the x-y plane is an *upward* force of 2 N?

Problem 4 Derive Eq. 3.20 using the same steps used to derive Eq. 3.15.

Problem 5 Assume a compressible cube with dimensions given in problem 1. Assume the cube has a mass of 2 kg. It is compressed holding the sides and bottom fixed until the top surface is as described in problem 2. (a) What is ρ_0 before and after the compression? (b) What is the actual density of the cube, $\rho = \frac{m}{V_{now}}$, before and after the deformation?

Problem 6 Consider the cube of problem 1 with no gravity acting. Given the following six forces applied to the three surfaces of the cube corresponding to the faces of the cube lying in the coordinate planes, compute the acceleration of the cube:

$$F_{13}(X_1, X_2, 2m) = -2N, \; F_{23}(X_1, X_2, 2m) = 0, \; F_{33}(X_1, X_2, 2m) = 0,$$
$$F_{12}(X_1, 2m, X_3) = +2N, \; F_{22}(X_1, 2m, X_3) = +0.5N, \; F_{23}(X_1, 2m, X_3) = 0,$$
$$F_{11}(2m, X_2, X_3) = -1N, \; F_{21}(2m, X_2, X_3) = 0, \; F_{31}(2m, X_2, X_3) = 2N,$$
$$F_{13}(X_1, X_2, 0) = +2N, \; F_{23}(X_1, X_2, 0) = 0, \; F_{33}(X_1, X_2, 0) = 0,$$
$$F_{12}(X_1, 0, X_3) = +1N, \; F_{22}(X_1, 0, X_3) = -0.5N, \; F_{23}(X_1, 0, X_3) = 0,$$
$$F_{11}(0, X_2, X_3) = -1N, \; F_{21}(0, X_2, X_3) = 0, \; F_{31}(0, X_2, X_3) = 2N.$$

Problem 7 Given the following distribution of $\mathcal{P}_{11}$, $\mathcal{P}_{21}$, and $\mathcal{P}_{31}$ on the uniform cube in problem 1 with a density $\rho_m = 1/4$kg, find the acceleration of the point (1m, 1m, 1m) inside the cube if there is no gravity.

$$\mathcal{P}_{11} = b x^2 - 2b x y + a z$$
$$\mathcal{P}_{21} = h x y^3 z$$
$$\mathcal{P}_{31} = d x^2 z$$
$$\mathcal{P}_{12} = d z^3$$

$$\mathcal{P}_{22} = a\,z - d y^2 x$$

$$\mathcal{P}_{32} = d x y z$$

$$\mathcal{P}_{13} = c y^2 - c x^2$$

$$\mathcal{P}_{23} = c x^2 + c y^2 + c z^2$$

$$\mathcal{P}_{33} = 0$$

with a = $1 N/m^2$, b = $1 N/m^3$, c = $1 N/m^4$, d = $1 N/m^5$, h = $1 N/m^7$.

Problem 8 If gravity is acting downward in the $-z$ direction, what is the acceleration of the point (1m, 1m, 1m) in problem 7?

Problem 9 What is the total force per unit volume on the top of the cube in problem 7?

Problem 10 What is the total force per unit initial volume in the x direction on the point (1,1,2) in problem 7?

Chapter 4
Force-Energy Relationships

This chapter describes the equation of motion found in the last chapter but this time in terms of energy. Since the last chapter described the equation of motion in terms of forces, to describe this relationship in terms of energy requires the relationship between force and energy. The familiar force-energy relationship of a simple spring motivates the surface-energy relationship for materials.

Springs

To describe the relationship between surface forces and energy, begin with a spring as described in an introductory physics class. When a spring is stretched an amount χ, the spring will exert a force F on the object pulling it, where

$$F = -k\chi, \qquad (4.1)$$

where "chi", χ, is the deformation of the spring. (Note that I have used "chi", χ, for the deformation instead of the usual "x".) The deformation χ can be described in terms of the length of the spring before the force is applied, L_0, and the length of the spring after the force is applied, L. This relationship is

$$\chi = L - L_0 \qquad (4.2)$$

The force, F, in Eq. 4.1 is the force exerted by the spring on whatever is pulling it. This force on the spring from whatever is pulling it is F_{pull}, where

---

Supplementary Information The online version contains supplementary material available at [https://doi.org/10.1007/978-3-031-09157-5_4].

Fig. 4.1 The upper picture
shows the spring at rest
(with no forces on it). The
lower picture shows the
spring displaced by χ from a
force, F_{pull}. The force from
the spring on the object
pulling it is F

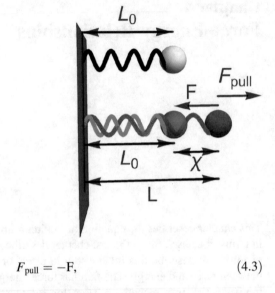

$$F_{pull} = -F, \qquad (4.3)$$

as illustrated in Fig. 4.1.

The constant k in Eq. 4.1 is called the spring constant, and it describes how
"strong" the spring is (i.e., a spring with a larger k requires more force to pull the
spring a given distance than a spring with a smaller k).

Young's Modulus

Consider two identical pieces of material, both described by a spring constant k. If
two pieces of this material are placed side by side, the model of the composite with a
single spring would require a spring constant of 2k. This is because the force
required to displace the composite material a given distance would now be twice
what it was before (Fig. 4.2). In general, the single spring constant must be a function
of the cross-sectional area of the material. The larger the cross-sectional area, the
larger the spring constant.

Similarly, if two pieces are placed end to end, the force required to displace the
material a given distance would be one half of the force required for the same
displacement on a single piece. This is because each piece of the composite material
would stretch one half of the total displacement (Fig. 4.3). In that case the spring
constant of the composite material would be $k/2$. In general, the spring constant is a
function of the length of the material. The longer the material, the smaller the spring
constant.

These effects are captured by defining the spring constant in terms of another
constant called Young's modulus, Y. Young's modulus, unlike the spring constant,
k, is only a function of the type of material, not the geometry. Using the initial cross-

Fig. 4.2 Figure showing
that the displacement force
is a function of the cross-
sectional area of the material
and thus so is the spring
constant

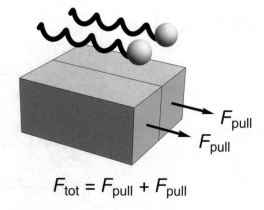

$$F_{tot} = F_{pull} + F_{pull}$$

Fig. 4.3 Displacement is
halved when material is
twice as long, resulting in
half the needed force and
therefore half the spring
constant to describe the
composite material

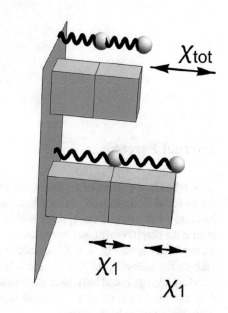

$$X_{tot} = X_1 + X_1$$

sectional area, A_0, and the initial length, L_0, of the material, the relationship between
Young's modulus and the spring constant is

$$k = A_0 Y / L_0. \tag{4.4}$$

Fig. 4.4. Modeling a
material with three springs

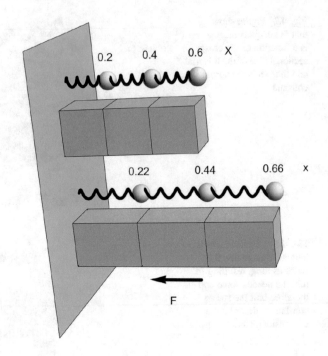

Internal Forces

Now let us turn to calculating the internal forces in the material. To do this, consider
the material as composed of many short springs as shown in Fig. 4.4. Represent the
location of the end of the springs within the material by the value X (capital "X")
before the displacement and the value of x (small "x") after the displacement. For the
example shown in Fig. 4.4, X will take on the values 0, 0.2, 0.4, and 0.6, and x will
take on the values 0, 0.22, 0.44, and 0.66 when the material has been stretched.

More springs could have been used with end points at 0, 0.1, 0.2, 0.3, 0.4, 0.5, and
0.6. The final position of these points would be 0, 0.11, 0.22, 0.33, 0.44, 0.55, and
0.66. Note that in both cases

$$x = 1.1^*X. \tag{4.5}$$

In fact, Eq. 4.5 applies no matter how many springs are used internally to
represent the material. Equation 4.5 is a mapping as described in Chap. 2. This
mapping takes every value of X into a new value, x. Note that no points are created or
destroyed and that every final value x has one and only one starting value, X. This is
the one-to-one mapping encountered in Chap. 2.

To find the internal forces within the material, consider one small spring somewhere within the model. Let ΔX be the initial length of that spring (L_0 in Eq. 4.2). Let
Δx be the final length of that spring (L in Eq. 4.2). The displacement of this spring, χ,
would be

$$\chi = \Delta x - \Delta X. \tag{4.6}$$

Combining Eqs. 4.4 and 4.6 with $L = \Delta x$ and $L_0 = \Delta X$ in Eq. 4.1,

$$F = -A_0 Y / (\Delta X)(\Delta x - \Delta X) = -A_0 Y \left(\frac{\Delta x}{\Delta X} - 1 \right) \tag{4.7}$$

and the internal force per unit initial area, f, would be

$$f = F/A_0 = -Y \left(\frac{\Delta x}{\Delta X} - 1 \right) \tag{4.8}$$

Finally letting the number of springs approach infinity while the length of each spring approaches zero, gives

$$f = - \lim_{X \to 0} Y \left(\frac{\Delta x}{\Delta X} - 1 \right) \tag{4.9}$$

which is

$$f = -Y \left(\frac{dx}{dX} - 1 \right). \tag{4.10}$$

In our example above, Eq. 4.10 with

$$dx/dX = 1.1 \tag{4.11}$$

yields

$$f = -Y(1.1 - 1) = -0.1Y \tag{4.12}$$

This is the force per unit initial area on the object pulling the material. The force per unit initial area on the material is

$$f = +0.1Y \tag{4.13}$$

Energy from Surface Forces

We now turn to finding the energy associated with the surface forces described in the last section. To do this let us again begin with a simple spring. Consider the energy of a spring, E, with a spring constant k that has been displaced by an amount χ,

$$E = 1/2k\chi^2, \tag{4.14}$$

with

$$\chi = \Delta - \Delta X = (x - x_0) - (X - X_0) \tag{4.15}$$

where

$x =$ current position of one end of the spring
$x_0 =$ current position of the other end of the spring
$X =$ initial position of one end of the spring
$X_0 =$ initial position of the other end of the spring.

The force from the end of the spring on the object pulling (or pushing) the spring is related to the energy of the spring, E as

$$F = -\frac{\partial E}{\partial x} \tag{4.16}$$

so that

$$F = -k\chi \tag{4.17}$$

as before.

Now let us consider a model with multiple springs. Following the pattern of the last section,

$$\chi = \Delta x - \Delta X \tag{4.18}$$

$$k = A_0 Y / L_0, \tag{4.19}$$

$$E = 1/2 \left(A_0 \frac{Y}{\Delta X} \right) (\Delta x - \Delta X)^2. \tag{4.20}$$

Divide this energy by the initial volume of the material, V_0, to get the energy per initial volume, ϵ, as

$$\epsilon = \frac{E}{V_0} = \frac{E}{\Delta X A_0} \tag{4.21}$$

Then, Eq. 4.21 gives

$$\epsilon = 1/2Y\left(\frac{\Delta x - \Delta X}{\Delta X}\right)^2. \tag{4.22}$$

Take the limit as ΔX approaches 0,

$$\epsilon = 1/2Y\left(\frac{dx}{dX} - 1\right)^2. \tag{4.23}$$

The force per unit initial area, f, is

$$f = F/A_0 = -\frac{d\epsilon}{d(dx/dX)}. \tag{4.24}$$

Substituting Eqs. 4.23 into 4.24 gives

$$f = -Y\left(\frac{dx}{dX} - 1\right), \tag{4.25}$$

which is the same as Eq. 4.10. The force per unit initial area on the material is

$$f = +Y\left(\frac{dx}{dX} - 1\right). \tag{4.26}$$

The previous example gave ϵ as a function of $\frac{dx}{dX}$ in one dimension and found the force per unit initial area in one dimension, f, by taking the derivative of ϵ with respect to $\frac{dx}{dX}$. We will assume that for two and three dimensions, ϵ is a function of the partial derivatives of x_i, i.e., $\frac{\partial x_i}{\partial X_j}$. This seems reasonable since the Taylor expansion of changes in x is to first order a function of $\frac{\partial x}{\partial X}$ as we found in Eq. 2.20. For the case where only one direction of the material is deformed at a time, the force per unit initial area on the material in the x, y, and z directions, f_{ii}, is

$$f_{ii} = +Y\left(\frac{\partial x_i}{\partial X_i} - 1\right), \quad i \text{ not summed} \tag{4.27}$$

$$\epsilon = 1/2Y\left(\frac{\partial x_i}{\partial X_i} - 1\right)^2, \quad i \text{ not summed} \tag{4.28}$$

$$f_{ii} = +\frac{\partial \epsilon}{\partial(\partial x_i/\partial X_i)}, \quad i \text{ not summed,} \tag{4.29}$$

Fig. 4.5 Deformation of a
region in the dx_2 direction,
with force applied on the A_2
plane

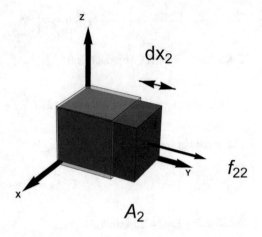

where f_{ii} is the ith component of the force per unit initial area in the ith direction,
where $i = 1$ is x, $i = 2$ is y, and $i = 3$ is z. (Fig. 4.5 shows the deformation for $i =$
2 and $f_{22} = \frac{F_{22}}{A_2}$.)

This example can be repeated for displacements parallel to the cube's surface to
motivate the relationship between f and ϵ. Let F_{21} be the force in the y direction on
the x plane. Then if the only change in the material is in the y direction, the force on
the material is

$$F_{21} = +k\chi, \tag{4.30}$$

with

$$\chi = \Delta x_2 \tag{4.31}$$

and

$$k = Y\frac{A_1}{\Delta X_1} \tag{4.32}$$

giving

$$\frac{F_{21}}{A_1} = +Y\frac{\Delta x_2}{\Delta X_1}. \tag{4.33}$$

In the limit as $X_1 \to 0$, f_{21} becomes

$$f_{21} = +Y\left(\frac{\partial x_2}{\partial X_1}\right), \tag{4.34}$$

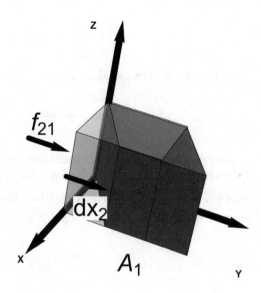

Fig. 4.6 Deformation of a region in the dx_2 direction, with force applied on the A_1 plane

so that f_{21} is the y-component of the force per unit initial area on the y-z plane, which is perpendicular to the x-axis, see Fig. 4.6. (Note that the y-z plane is denoted with 1 corresponding to the x direction.)

This can be repeated in each direction to give

$$f_{ij} = Y \frac{\partial x_i}{\partial X_1}, \text{ with } i \neq j, \tag{4.35}$$

where f_{ij} is the ith component of the force per unit initial area acting on the jth plane.
Using

$$E = 1/2 \, k \chi^2, \tag{4.36}$$

the energy associated with F_{21} is

$$E = 1/2 \, Y \, \frac{A_1}{\Delta X_1} (\Delta x_2)^2. \tag{4.37}$$

Again, writing the energy per initial volume gives

$$\epsilon = \frac{E}{V_0} = \frac{E}{A_1 \Delta X_1} = \lim_{\Delta X_1 \to 0} 1/2Y \left(\frac{\Delta x_2}{\Delta X_1} \right)^2 = 1/2Y \left(\frac{\partial x_2}{\partial X_1} \right)^2 \tag{4.38}$$

and

$$f_{21} = \frac{\partial \epsilon}{\partial(\partial x_2/\partial X_1)}. \tag{4.39}$$

Repeating this in each direction gives

$$f_{ij} = \frac{\partial \epsilon}{\partial(\partial x_i/\partial X_j)}, \quad \text{with } i \neq j. \tag{4.40}$$

An example of the force per initial area on the x face in the y direction, f_{21}, is shown in Fig. 4.6. Equations 4.29 and 4.34 were made assuming ϵ is a function of only $\left(\frac{\partial x_i}{\partial X_j}\right)^2$. In the more general case, ϵ may be any function of $\frac{\partial x_i}{\partial X_j}$. When this is the case, we can identify the force per initial area as engineering stress, $\mathcal{P}$,

$$\mathcal{P}_{ij} = \frac{\partial \epsilon}{\partial(\partial x_i/\partial X_j)} \tag{4.41}$$

with ϵ any function of the nine values, $\partial x_i/\partial X_j$.

A more general derivation of Eq. 4.41 can be made using the n-dimensional divergence theorem and the Euler-Lagrange formulation of stress and strain (see Chap. 14), but here we simply state Eq. 4.41 as an hypothesis to be tested by later examples. In the next chapter we will find restrictions on which combinations of $\frac{\partial x_i}{\partial X_j}$ are physically possible depending upon the properties of the material of interest.

—————————— Aside ——————————

The relationship

$$f = -\frac{d\epsilon_{\text{surf}}}{d(dx/dX)} \tag{4.42}$$

may look a bit peculiar since it is a derivative with respect to a derivative, but if dx/dX is defined as a variable, say r, then the derivative would be

$$-\frac{d\epsilon_{\text{surf}}}{d(dx/dX)} = -\frac{d\epsilon_{\text{surf}}}{dr} = -\frac{d}{dr}\left(1/2 Y(r-1)^2\right) = -Y(r-1) \tag{4.43}$$

which when r is replaced by dx/dX is

$$-Y(dx/dX - 1) \tag{4.44}$$

which according to Eq. 4.26 is indeed f.

———————————————————————————————

Equation of Motion in Terms of Energy

The surface and body forces can now be described in terms of energy for an isotropic material. Substituting Eq. 4.41 into Eq. 3.16,

$$\rho_0 \, a_j = \frac{\partial}{\partial X_i} \left(\frac{\partial \epsilon}{\partial (\partial x_j / \partial X_i)} \right) - \rho_0 \, g \, \delta_{j3}. \qquad (4.45)$$

What remains is to find ϵ for a particular material. This must be done experimentally as described in Chaps. 10 and 12, but first we will consider constraints on ϵ for isotropic bodies in Chap. 5.

Problems

Problem 1 Find the force needed to stretch a cuboid of material by 0.1m if the only force is in the L_0 direction if $A_0 = 2 \; m^2$, $L_0 = 1m$, and $Y = 4 \; N/m^2$.

Problem 2 Find pulling force within the material at x $= 0.22$ in Fig. 4.4 if the original dimensions of the entire bar are $A_0 = 2 \; m^2$, $L_0 = 1m$, and $Y = 4 \; N/m^2$ by (a) using Eq. 4.1 and (b) using Eq. 4.10.

Problem 3 If $\epsilon = \left(\frac{\partial x_1}{\partial X_1} \right)^2 + \left(\frac{\partial x_2}{\partial X_2} \right)^2 + \left(\frac{\partial x_3}{\partial X_3} \right)^2$, find $\mathcal{P}_{ij}$.

Problem 4 Given a rectangular bar of dimensions 2 m, 2 m, and 6 m, let the 2 m dimensions lie along the y and z coordinate directions and the 6 m dimension lie along the x coordinate direction. Place one corner of the bar at (0, 0, 0). Given $\epsilon = 5N/m^2 \left(\left(\frac{\partial x_1}{\partial x_1} \right)^3 - 1 \right)$, find the change in energy of the bar if it is stretched to 2 m, 2 m, and 9 m.

Problem 5 Find the force on the long end of the bar in problem 4.

Problem 6 If $\epsilon = \left(\frac{\partial x_1}{\partial X_1} \right)^2 + \left(\frac{\partial x_2}{\partial X_2} \right)^2 + \left(\frac{\partial x_3}{\partial X_3} \right)^2$, write the three equations of motion of any part of the material.

Problem 7 If $\epsilon = 0.5N/m^2 \left(\frac{\partial x_1}{\partial x_1} \frac{\partial x_2}{\partial x_2} \frac{\partial x_3}{\partial x_3} - 1 \right)$, find the change in energy of the bar in problem 4 if the deformation of the bar is described as the mapping,

$$x_1 = (0.2/m)X_1^2 + 0.2X_2 + 3m$$
$$x_2 = 1.3X_1 + 1.5X_2 + X_3 - 1m$$
$$x_3 = X_3.$$

Problem 8 Find the final force on the end of the bar in problem 7 which was originally at $X_1 = 6$ m. If you have graphics software, graph the final configuration of the bar.

Chapter 5
Isotropic Materials

In the last chapter, we have hypothesized that the energy that has been rescaled by dividing it by the initial volume of the material is a function of $\frac{\partial x_i}{\partial X_j}$. In this chapter, we will find that there are constraints on which functions of $\frac{\partial x_i}{\partial X_j}$ are physically possible. We will divide materials into two groups – isotropic and anisotropic. Isotropic materials may be cut along any direction and placed in service with no change in its response to force. Anisotropic materials have a grain or an embedded structure that produce a different response to force when it is placed in different orientations to the force. Both isotropic and anisotropic materials can be rotated after a deformation without changing the energy of deformation. Only isotropic materials can undergo a general rotation before the deformation with no change in the energy of deformation. Both anisotropic and anisotropic materials can be translated with no change in the energy of deformation. (The energy associated with body forces may change – as the gravitational energy changes with position, but a change in the deformation energy requires a deformation of the material.) Examples of isotropic materials would be most rubbers, polymers, and metals. Examples of anisotropic materials would be wood with prominent grain, concrete with re-barb, many fabrics, and even paper. We will consider isotropic materials here. Anisotropic materials will be covered in Chap. 12.

Rotations and Translations

In Chap. 2, we defined an affine mapping to be a mapping from the points separated by dX_j to the same points separated by dx_i. The affine mapping in 3D is (Eq. 2.24)

© The Author(s), under exclusive license to Springer Nature Switzerland AG 2022
H. Hardy, *Engineering Elasticity*, https://doi.org/10.1007/978-3-031-09157-5_5

$$dx_i = \frac{\partial x_i}{\partial X_j} dX_j. \tag{5.1}$$

Let us see what happens to $\frac{\partial x_i}{\partial X_j}$ if we rotate the body after the affine mapping. A general rotation in 3D can be accomplished by applying a rotation matrix to the final point locations. A general rotation matrix in 3D, $\mathcal{R}$, is

$$\mathcal{R} = \mathcal{R}_x \mathcal{R}_y \mathcal{R}_z, \tag{5.2}$$

where

$$\mathcal{R}_x = \begin{pmatrix} 1 & 0 & 0 \\ 0 & \cos\alpha & -\sin\alpha \\ 0 & \sin\alpha & \cos\alpha \end{pmatrix}$$

$$\mathcal{R}_y = \begin{pmatrix} \cos\beta & 0 & \sin\beta \\ 0 & 1 & 0 \\ -\sin\beta & 0 & \cos\beta \end{pmatrix} \tag{5.3}$$

$$\mathcal{R}_z = \begin{pmatrix} \cos\gamma & -\sin\gamma & 0 \\ \sin\gamma & \cos\gamma & 0 \\ 0 & 0 & 1 \end{pmatrix}$$

with α, β, and γ being the angle of rotation of the body about the x, y, and z coordinate axes, respectively. Thus, any point in the body before a rotation, x_j, will become x_i after the rotation, where

$$\bar{x}_i = \mathcal{R}_{ij} x_j. \tag{5.4}$$

A translation of the point, x_i, will add a vector to the point position

$$x_i = x_i + v_i. \tag{5.5}$$

We are interested in what happens to $\frac{\partial x_i}{\partial X_i}$ if a rotation and translation occurs.

First, we will see that dx_i and dX_i in Eq. 5.1 are unchanged by a translation. If $x_i^{(1)}$ are the three components of one point and $x_i^{(2)}$ are the three components of another point, then

$$dx_i = x_i^{(2)} - x_i^{(1)}. \tag{5.6}$$

If we apply the same translation to each point:

$$dx_i = \left(x_i^{(2)} + v_j\right) - \left(x_i^{(1)} + v_j\right). \tag{5.7}$$

The v_i term subtracts out and dx_i is unchanged. The same derivation can be carried out for X_i showing that dX_i is also not changed by a translation. From this, we conclude that neither a translation before nor after the deformation will change the terms in Eq. 5.1.

Unfortunately, this is not the case for rotations. First, we will rotate the points after the deformation. Equation 5.4 gives the new point locations. Applying a rotation as $\mathcal{R}_{ik}$ to each of the final points, x_i,

$$d\bar{x}_k = \mathcal{R}_{ik} x_k^{(2)} - \mathcal{R}_{ik} x_k^{(1)} = \mathcal{R}_{ik} dx_k, \tag{5.8}$$

where $d\bar{x}_k$ is the separation of the points after the rotation. Using Eq. 5.1, we have

$$d\bar{x}_i = \mathcal{R}_{ik} \frac{\partial x_k}{\partial X_j} dX_j, \tag{5.9}$$

so that the new affine mapping is

$$\frac{\partial \bar{x}_i}{\partial X_j} = \mathcal{R}_{ik} \frac{\partial x_k}{\partial X_j}. \tag{5.10}$$

and the values of the matrix $\frac{\partial x_i}{\partial X_j}$ have changed with the rotation, $\mathcal{R}$.

If we rotate the body before the rotation, then we rotate the dX_i vectors:

$$d\bar{X}_i = \mathcal{R}_{ik} dX_i \tag{5.11}$$

and with $d\bar{X}_k$ resulting from a rotation before the deformation,

$$\frac{\partial \bar{x}_k}{\partial X_i} = \frac{\partial x_k}{\partial X_j} \mathcal{R}_{ji} dX_i, \tag{5.12}$$

so that once again, the values of $\frac{\partial x_k}{\partial X_j}$ have been changed and

$$\frac{\partial \bar{x}_k}{\partial X_i} = \frac{\partial x_k}{\partial X_j} \mathcal{R}_{ji}. \tag{5.13}$$

For the most general case of rotations before and after a deformation,

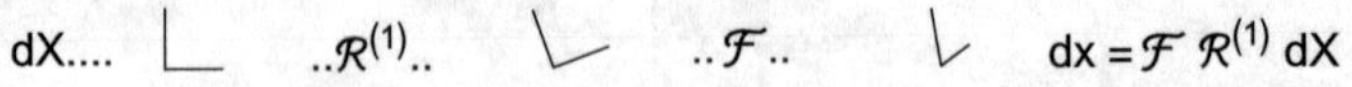

$$dX.... \quad \llcorner \quad ..\mathcal{R}^{(1)}.. \quad \diagdown \quad ..\mathcal{F}.. \quad \diagdown \quad dx = \mathcal{F}\,\mathcal{R}^{(1)}\,dX$$

Fig. 5.1 From left to right. Begin with dX (x-y coordinate axes). Rotate material, $\mathcal{R}^{(1)}$ (by 20^0 clockwise). Then apply a deformation $\mathcal{F}$ (a 20% compression horizontally). The combined deformation of dX is then $dx = \mathcal{F}\mathcal{R}^{(1)}dX$

$$dX.... \quad \llcorner \quad ..\mathcal{F}.. \quad \llcorner \quad ..\mathcal{R}^{(2)}.. \quad \diagdown \quad dx = \mathcal{R}^{(2)}\,\mathcal{F}\,dX$$

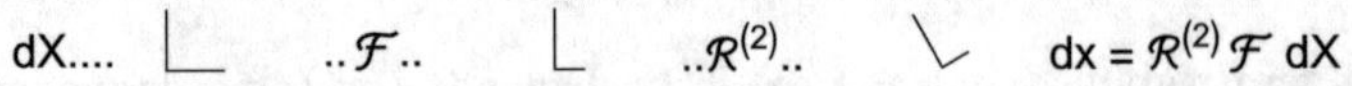

Fig. 5.2 From left to right. Begin with dX (x-y coordinate axes). Deform the material, $\mathcal{F}$ (a 20% compression horizontally). Then rotate material, $\mathcal{R}^{(2)}$ (by 30^0 clockwise). The combined deformation of dX is then $dx = \mathcal{F}\mathcal{R}^{(1)}dX$

$$\frac{\partial \bar{x}_i}{\partial X_m} = \mathcal{R}^{(2)}{}_{ik}\,\frac{\partial x_k}{\partial X_j}\,\mathcal{R}^{(1)}{}_{jm},\tag{5.14}$$

where the rotation before, $\mathcal{R}^{(1)}$, is not necessarily the same as the rotation afterward, $\mathcal{R}^{(2)}$.

A couple of pictures may help. Figures 5.1 and 5.2 illustrate the effect of rotations on the deformation gradient matrix, $\mathcal{F}$.

For isotropic bodies neither rotation before nor after the deformation should change ϵ.

Rotational Invariants

If a material is isotropic, ϵ must not change in the mapping defined by Eq. 5.14. Thus, we must build $\epsilon\left(\frac{\partial x_i}{\partial X_j}\right)$ with combinations of $\frac{\partial x_i}{\partial X_j}$ that are unchanged with the rotation of the body.

Just as we expressed $\mathcal{R}_{ij}$ in matrix notation, Eq. 5.3, we can also express $\frac{\partial x_i}{\partial X_j}$ and $\frac{\partial \bar{x}_i}{\partial X_j}$ in matrix notation as follows:

$$\mathcal{F} = \mathcal{F}_{ij} = \frac{\partial x_i}{\partial X_j} = \begin{pmatrix} \dfrac{\partial x_1}{\partial X_1} & \dfrac{\partial x_1}{\partial X_2} & \dfrac{\partial x_1}{\partial X_3} \\[2mm] \dfrac{\partial x_2}{\partial X_1} & \dfrac{\partial x_2}{\partial X_2} & \dfrac{\partial x_2}{\partial X_3} \\[2mm] \dfrac{\partial x_3}{\partial X_1} & \dfrac{\partial x_3}{\partial X_2} & \dfrac{\partial x_3}{\partial X_3} \end{pmatrix}\tag{5.15}$$

and

$$\mathcal{F} = \mathcal{F}_{ij} = \frac{\partial \bar{x}_i}{\partial X_j} = \begin{pmatrix} \dfrac{\partial \bar{x}_1}{\partial X_1} & \dfrac{\partial \bar{x}_1}{\partial X_2} & \dfrac{\partial \bar{x}_1}{\partial X_3} \\[2mm] \dfrac{\partial \bar{x}_2}{\partial X_1} & \dfrac{\partial \bar{x}_2}{\partial X_2} & \dfrac{\partial \bar{x}_2}{\partial X_3} \\[2mm] \dfrac{\partial \bar{x}_3}{\partial X_1} & \dfrac{\partial \bar{x}_3}{\partial X_2} & \dfrac{\partial \bar{x}_3}{\partial X_3} \end{pmatrix}, \tag{5.16}$$

then Eq. 5.14 can be written as

$$\mathcal{F} = \mathcal{R}^{(1)} \mathcal{F} \mathcal{R}^{(2)}. \tag{5.17}$$

What combinations of the $\frac{\partial x_i}{\partial X_j}$ values should be expected to be invariant to rotations? We learned in Chap. 2 that the coordinate vectors $\hat{i}, \hat{j}$, and $\hat{k}$ map locally into the column vectors of $\mathcal{F}$

$$\begin{aligned} \vec{a} &= \left(\frac{\partial x_1}{\partial X_1}, \frac{\partial x_2}{\partial X_1}, \frac{\partial x_3}{\partial X_1} \right) \\ \vec{b} &= \left(\frac{\partial x_1}{\partial X_2}, \frac{\partial x_2}{\partial X_2}, \frac{\partial x_3}{\partial X_2} \right) \\ \vec{c} &= \left(\frac{\partial x_1}{\partial X_3}, \frac{\partial x_2}{\partial X_3}, \frac{\partial x_3}{\partial X_3} \right). \end{aligned} \tag{5.18}$$

Thus, if we inscribe $\hat{i}, \hat{j}$, and $\hat{k}$ vectors onto our initial body, $\vec{a}, \vec{b}$, and $\vec{c}$ will be the locations of these vectors after the deformation $\frac{\partial x_i}{\partial X_j}$. We know from the properties of vectors that dot products and the magnitude of cross products of vectors are invariant under coordinate rotations. Thus, we would expect some combination of dot and cross products of $\vec{a}, \vec{b}$, and $\vec{c}$ to be invariant. Also, we can rotate the body prior to deformation so that the $\hat{i}, \hat{j}$, and $\hat{k}$ vectors in the body becomes the $\hat{j}, \hat{k}$, and $\hat{i}$ or the $\hat{k}, \hat{j}$, and $\hat{i}$ vectors, respectively. Thus, the invariant also needs to be unchanged by a cyclic permutation of $\vec{a}, \vec{b}$, and $\vec{c}$. Of course, the final form of the invariants must be validated by direct computation as done in problems 8 and 9, but the result is that the invariants needed are I_1, I_2, and I_3 defined as

$$\begin{aligned} \mathcal{I}_1 &= \vec{a}.\vec{a} + \vec{b}.\vec{b} + \vec{c}.\vec{c} \\ \mathcal{I}_2 &= \left(\vec{a} \times \vec{b} \right).\left(\vec{a} \times \vec{b} \right) + \left(\vec{a} \times \vec{c} \right).\left(\vec{a} \times \vec{c} \right) + \left(\vec{b} \times \vec{c} \right).\left(\vec{b} \times \vec{c} \right) \\ \mathcal{I}_3 &= \vec{a}.\left(\vec{b} \times \vec{c} \right), \end{aligned} \tag{5.19}$$

where $\vec{a}, \vec{b}$, and $\vec{c}$ are the column vectors of $\mathcal{F}_{ij}$, i.e., Eq. 5.18.

Since I_1, I_2, and I_3 are invariant to rotations of the body, any function of I_1, I_2, and I_3 will also be invariant. This means that proper energy functions of isotropic bodies can be any function of I_1, I_2, and I_3. That is,

$$\epsilon = f(\mathcal{I}_1, \mathcal{I}_2, \mathcal{I}_3). \tag{5.20}$$

Expanding Eq. 5.19, we have

$$\mathcal{I}_1 = \left(\frac{\partial x_1}{\partial X_1}\right)^2 + \left(\frac{\partial x_2}{\partial X_1}\right)^2 + \left(\frac{\partial x_3}{\partial X_1}\right)^2 + \left(\frac{\partial x_1}{\partial X_2}\right)^2 + \left(\frac{\partial x_2}{\partial X_2}\right)^2 + \left(\frac{\partial x_3}{\partial X_2}\right)^2$$
$$+ \left(\frac{\partial x_1}{\partial X_3}\right)^2 + \left(\frac{\partial x_2}{\partial X_3}\right)^2 + \left(\frac{\partial x_3}{\partial X_3}\right)^2$$

$$\mathcal{I}_2 = \left(\frac{\partial x_1}{\partial X_1}\frac{\partial x_2}{\partial X_2} - \frac{\partial x_1}{\partial X_2}\frac{\partial x_2}{\partial X_1}\right)^2 + \left(\frac{\partial x_1}{\partial X_3}\frac{\partial x_2}{\partial X_1} - \frac{\partial x_1}{\partial X_1}\frac{\partial x_2}{\partial X_3}\right)^2$$
$$+ \left(\frac{\partial x_1}{\partial X_2}\frac{\partial x_2}{\partial X_3} - \frac{\partial x_1}{\partial X_3}\frac{\partial x_2}{\partial X_2}\right)^2 + \left(\frac{\partial x_1}{\partial X_1}\frac{\partial x_3}{\partial X_2} - \frac{\partial x_1}{\partial X_2}\frac{\partial x_3}{\partial X_1}\right)^2$$
$$+ \left(\frac{\partial x_2}{\partial X_1}\frac{\partial x_3}{\partial X_2} - \frac{\partial x_2}{\partial X_2}\frac{\partial x_3}{\partial X_1}\right)^2 + \left(\frac{\partial x_1}{\partial X_3}\frac{\partial x_3}{\partial X_1} - \frac{\partial x_1}{\partial X_1}\frac{\partial x_3}{\partial X_3}\right)^2 \tag{5.21}$$
$$+ \left(\frac{\partial x_1}{\partial X_2}\frac{\partial x_3}{\partial X_3} - \frac{\partial x_1}{\partial X_3}\frac{\partial x_3}{\partial X_2}\right)^2 + \left(\frac{\partial x_2}{\partial X_3}\frac{\partial x_3}{\partial X_1} - \frac{\partial x_2}{\partial X_1}\frac{\partial x_3}{\partial X_3}\right)^2$$
$$+ \left(\frac{\partial x_2}{\partial X_2}\frac{\partial x_3}{\partial X_3} - \frac{\partial x_2}{\partial X_3}\frac{\partial x_3}{\partial X_2}\right)^2$$

$$\mathcal{I}_3 = \frac{\partial x_1}{\partial X_1}\left(\frac{\partial x_2}{\partial X2}\frac{\partial x_3}{\partial X_3} - \frac{\partial x_2}{\partial X_3}\frac{\partial x_3}{\partial X_2}\right) + \frac{\partial x_2}{\partial X_1}\left(\frac{\partial x_1}{\partial X3}\frac{\partial x_3}{\partial X_2} - \frac{\partial x_1}{\partial X_2}\frac{\partial x_3}{\partial X_3}\right)$$
$$+ \frac{\partial x_3}{\partial X_1}\left(\frac{\partial x_1}{\partial X2}\frac{\partial x_2}{\partial X_3} - \frac{\partial x_1}{\partial X_3}\frac{\partial x_2}{\partial X_2}\right).$$

In conclusion, ϵ for isotropic bodies must be expressed as a function of the rotational invariants I_1, I_2, and I_3 given in Eq. 5.21.

The I_i functions are quite complicated functions of $\frac{\partial x_i}{\partial X_j}$, and the I_i values are nonlinear. The complexity in the I_i functions is "built in" to the physics of large deformations or rotations of isotropic bodies. Chap. 10 shows how to experimentally determine the functional form of ϵ for isotropic bodies. Chapter 12 finds the invariants for anisotropic bodies. In Chaps. 7 and 8, we will discover how to use several functional forms of $\epsilon(I_i)$ to simulate the deformation of a body under a number of conditions. In Chap. 6, we will discuss the application of Eq. 5.21 to deformations that are slow compared to the response time of the materials.

Problems

Problem 1 Consider the deformation

$$x_1 = 2X_1 + 3X_2 + X_3 + 2$$
$$x_2 = 1X_1 - 3X_2 + 2X_3 + 1$$
$$x_3 = 0.5X_1 - 1X_2 - 0.5X_3 + 2.$$

Find the $\mathcal{F}$ matrix for this deformation.

Problem 2 Consider the deformation $\mathcal{F} = \begin{pmatrix} .8 & .2 & -.3 \\ -.1 & 1.1 & .2 \\ .3 & .5 & 0.9 \end{pmatrix}$. Show that $\mathcal{F}$ is

unchanged by a translation of $(1,3,1)$. (Optional: Graph a $1 \times 1 \times 1$ cube of the material with a center located at $(1, 1, 1)$ before the deformation, the same cube after the deformation and translation.)

Problem 3 Consider the deformation

$$x_1 = .8X_1 + .2X_2 - .3X_3 + 1$$
$$x_2 = -.1X_1 + 1.1X_2 + 2X_3 + 3$$
$$x_3 = 0.3X_1 + .5X_2 + 0.9X_3 + 1.$$

Calculate I_1, I_2, and I_3 for that deformation.

Problem 4 Consider the simple deformation $\mathcal{F} = \begin{pmatrix} 2 & 0 & 0 \\ 0 & 1 & 0 \\ 0 & 0 & 1 \end{pmatrix}$ of a unit cube with

corner points $\{0, 0, 0\}$, $\{1, 0, 0\}$, $\{0, 1, 0\}$, and $\{0, 0, 1\}$. (a) Find the location of these point after the deformation $\mathcal{F}$. (b) Find the location of these points if the cube is first rotated by 20^0 about the z-axis and then deformed by $\mathcal{F}$. (c) Find the location of these point if the cube is first deformed by $\mathcal{F}$ and then rotated by 20^0 about the z-axis. (d) If you have plotting software, draw the initial and final cube for cases a, b, and c.

Problem 5 Show that $\mathcal{R}$ with $\alpha = 0^0$, $\beta = 90^0$, and $\gamma = 90^0$ rotates $\widehat{i}, \widehat{j}$, and $\widehat{k}$ into $\widehat{j}, \widehat{k}$, and $\widehat{i}$, respectively, and that $\mathcal{R}$ with $\alpha = 270^0$, $\beta = 360^0$, and $\gamma = 270^0$ rotates $\widehat{i}, \widehat{j}$, and $\widehat{k}$ into $\widehat{k}, \widehat{i}$, and $\widehat{j}$, respectively.

Problem 6 For the deformation $\mathcal{F} = \begin{pmatrix} .8 & .2 & -.3 \\ -.1 & 1.1 & .2 \\ .3 & .5 & 0.9 \end{pmatrix}$ followed by a rotation,

$\mathcal{R} = \mathcal{R} = \begin{pmatrix} 1 & 0 & 0 \\ 0 & \cos(20^0) & -\sin(20^0) \\ 0 & \sin(20^0) & \cos(20^0) \end{pmatrix}$, show that the composite deformation

gives the same I_1 as the I_1 for the original deformation.

Problem 7 The first deformation is a rotation, $\mathcal{R} = \begin{pmatrix} 1 & 0 & 0 \\ 0 & \cos(20^0) & -\sin(20^0) \\ 0 & \sin(20^0) & \cos(20^0) \end{pmatrix}$.

The second is the deformation $\mathcal{F} = \begin{pmatrix} .8 & .2 & -.3 \\ -.1 & 1.1 & .2 \\ .3 & .5 & 0.9 \end{pmatrix}$. Show that the I_3 for the

second deformation is the same as for the combination deformation of the rotation followed by the deformation.

Problem 8 If you have access to an algebraic solver, show that I_1, I_2, and I_3 are all unchanged by a rotation after the deformation. If not show it for the deformation

$\mathcal{F} = \begin{pmatrix} 1.2 & .2 & .5 \\ -.3 & .8 & .1 \\ .6 & -.2 & 1.5 \end{pmatrix}$, $\mathcal{R}^{(2)} = \begin{pmatrix} 1 & 0 & 0 \\ 0 & \cos(20^0) & -\sin(20^0) \\ 0 & \sin(20^0) & \cos(20^0) \end{pmatrix}$.

Problem 9 If you have access to an algebraic solver, show that I_1, I_2, and I_3 are all unchanged by a rotation before the deformation. If not, show it for the deformation

$\mathcal{F} = \begin{pmatrix} 1.2 & .2 & .5 \\ -.3 & .8 & .1 \\ .6 & -.2 & 1.5 \end{pmatrix}$, $\mathcal{R}^{(2)} = \begin{pmatrix} 1 & 0 & 0 \\ 0 & \cos(20^0) & -\sin(20^0) \\ 0 & \sin(20^0) & \cos(20^0) \end{pmatrix}$.

Problem 10 Find the acceleration of the point that was originally at $(2, 1, 4)$ if the deformation is defined by Eq. 2.9 with $\epsilon = I_1$ and $\rho_0 = 2$. (Ignore units for this problem.)

Problem 11 If you have access to a 3D graphics routine, consider a unit cube with its corners on $(0, 0, 0)$ and $(1, 1, 1)$ with its edges lying on the coordinate axes. Plot this same cube before and after a rotation of 20^0 about the z-axis.

Chapter 6
Minimizing Energy

This chapter will focus on solving finite deformations when acceleration can be ignored. This type of solution is important when deformations are slow compared to the relaxation time it takes for a material to respond to external forces. This process is called a quasi-static deformation. To solve this problem, we will solve Eq. 4.45 when all components of the acceleration, a_j, are taken to be zero. Of course, a_j is never zero in a displacement, but when displacements are performed slowly, it is a satisfactory approximation. The only issue is that the nonlinear equations describing the deformation may have multiple solutions. To find the correct physical solution, there must be a continuous path from initial to final deformation. This is simulated by dividing any applied force or displacement into small steps and finding the quasi-static solution for each step.

In the body of this text, we will solve for the deformation of the body by finding the final configuration of the material which minimizes the energy of the material. An alternative approach is to substitute a function form of $\epsilon \left(\frac{\partial x_i}{\partial X_j} \right)$ into Eq. 4.45 and use finite elements to solve for the final configuration of the material. This latter approach is taken in Appendix D. The minimization of energy approach is described in great detail within the body of the book, but not the finite element solution technique, because finite element solution techniques are well described in many other texts. Showing that both techniques provide the same results to an example problem provides some assurance that the equations have been solved correctly by both techniques. The examples in Appendix D and Chaps. 7, 8, and 12 are quasi-static solutions of Eq. 4.45. Chapter 11 provides a time-dependent example.

(Notational warning: For this chapter only, x_i is used to be the x component of the ith node, not the x, y, and z components of any point in the body as in other chapters)

Supplementary Information The online version contains supplementary material available at [https://doi.org/10.1007/978-3-031-09157-5_6].

Spring Model

A calculation similar to the one in Chap. 4 is to be carried out, but this time assuming the final positions of the nodes are unknown. To keep things simple, our example has only three nodes as shown in Fig. 6.1. We wish to find the final position of these nodes in a quasi-static process with a force of magnitude F_{app} in the $-x$ direction applied to the end of the spring, so as to compress the two springs. This type of problem is called a "mixed" boundary condition problem since both Neumann and Dirichlet boundary conditions are involved. Neumann boundary conditions fix the derivative of the position of a surface node, here defined by the applied force, $\vec{F}_{app}$, and Dirichlet boundary conditions fix surface points, here the node at which $(0, 0)$ is fixed.

The initial position of the ith node is X_i and the final position of the ith node is x_i. To keep the equations from being too messy, units will be dropped for this calculation. Consider a specific case of three nodes. Assume the initial positions of all the nodes are as follows:

$$
\begin{aligned}
x_1 &= 0 \\
x_2 &= 1 \\
x_3 &= 2.
\end{aligned}
\tag{6.1}
$$

Apply a force, $F_{app} = 0.2$, toward the wall. Node 1 is fixed, so its final position is known, but the final positions of nodes 2 and 3 are not known, i.e.,

Fig. 6.1 Simple spring model with three nodes

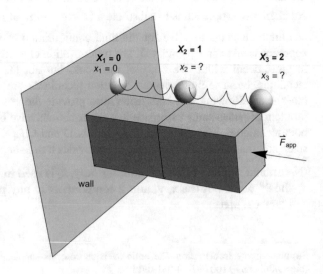

$$x_1 = 0$$
$$x_2 = ?$$
$$x_3 = ?$$
(6.2)

The physics of the problem dictates that for each node

$$\sum F = 0.$$
(6.3)

This is a one-dimensional problem, so the forces only have components in the x direction. Also note that the force of each spring on the node to the left of the spring is negative and the force of each spring on the node to the right of the spring is positive.

For the center node, node 2,

$$F_{\text{left spring}} = k(|x_2 - x_1| - |X_2 - X_1|)$$
(6.4)

$$F_{\text{right spring}} = -k(|x_3 - x_2| - |X_3 - X_2|).$$
(6.5)

Note that the absolute values can be dropped because x_i is always greater than x_{i-1} and X_i is always greater than X_{i-1} (i.e., the material will never be inverted.) The sum of the forces (Eq. 6.3) on node 2 is

$$k((x_2 - x_1) - (X_2 - X_1)) - k((x_3 - x_2) - (X_3 - X_2)) = 0.$$
(6.6)

For node 3,

$$F_{\text{left spring}} = -k((x_3 - x_2) - (X_3 - X_2)).$$
(6.7)

There is no spring to the right of node 3, but there is an applied force there of magnitude F_{app}, and the sum of the force on node 3 is

$$-k((x_3 - x_2) - (X_3 - X_2)) - F_{\text{app}} = 0.$$
(6.8)

The final position of node 1 is known, so there is no need to set up Eq. 6.3 for node 1 to find the final position of node 1.

There are two unknowns, x_2 and x_3, and two linear equations, Eqs. 6.6 and 6.8. Solving Eqs. 6.6 and 6.8 for x_2 and x_3 using a spring constant, k, of 1 gives

$$x_2 = 0.7$$
$$x_3 = 1.4.$$
(6.9)

We have found the final unknown positions of nodes 2 and 3 as we desired.

Now let us approach this problem another way. Instead of setting up the sum of the forces on each node, let us find when the total energy of the system, E_{tot}, is a minimum. We will find that the total energy that must be minimized is the total energy stored in the springs plus the energy added to the system by the applied force. That is,

$$E_{tot} = \sum_{i=1}^{N} E_j - W_{app}, \qquad (6.10)$$

where

E_j = energy of spring j

W_{app}= work done by the applied force on the surface node = $\int_{X_3}^{x_3} \vec{F}_{app} \circ d\vec{x}$

To show that minimizing E_{tot} gives the same equations as the more familiar $\sum F$ =0, let us solve the same problem stated above but this time by minimizing E_{tot}. From Eqs. 4.14 and 4.15,

$$E_i = 1/2 \, k \left((x_i - x_{i-1}) - (X_i - X_{i-1})\right)^2, \qquad (6.11)$$

for $i = 1, 2,$ and 3. Using that the applied force is constant, $\vec{F}_{app} = (-F_{app}, 0, 0)$ and $\vec{x}_3 - \vec{X}_3 = (x_3 - X_3, 0, 0)$. We have

$$W_{app} = \int_{X_3}^{x_3} \vec{F}_{app} \circ d\vec{s} =$$
$$\vec{F}_{app} \circ \int_{X_3}^{x_3} d\vec{x} = (-F_{app}, 0, 0) \circ (x_3 - X_3, 0, 0) \qquad (6.12)$$

and

$$W_{app} = -F_{app}(x_3 - X_3). \qquad (6.13)$$

Using Eqs. 6.11 and 6.13 in Eq. 6.10 gives

$$E_{tot} = 1/2 \, k((x_2 - x_1) - (X_2 - X_1))^2 + 1/2 \, k \left((x_3 - x_2) - (X_3 - X_2)\right)^2 \\ - \left(-F_{app}(x_3 - X_3)\right). \qquad (6.14)$$

Now to minimize E_{tot}, vary the values of x_2 and x_3, and find when $\frac{\partial E_{tot}}{\partial x_2}$ and $\frac{\partial E_{tot}}{\partial x_3}$ are both equal zero. That is,

$$\frac{\partial E_{tot}}{\partial x_2} = k((x_2 - x_1) - (X_2 - X_1)) - k((x_3 - x_2) - (X_3 - X_2)) = 0 \qquad (6.15)$$

and

$$\frac{\partial E_{tot}}{\partial x_3} = k((x_3 - x_2) - (X_3 - X_2)) + F_{app} = 0. \tag{6.16}$$

Notice that these are the same two equations that were found from $\Sigma F = 0$, i.e., Eqs. 6.6 and 6.8! Therefore, the final solution using minimum energy will be the same as the solution from the sum of forces.

Discrete 3D Model

To use the same minimizing energy approach for 3D that we used for the springs in the last section, we need to express the deformation energy in terms of 3D volumes, add gravity to the equations, and express the external forces as forces against surfaces. These changes expand Eq. 6.10 to

$$E_{tot} = \sum_{i=1}^{N_v} E_i - \sum_{k=1}^{N_n} \left(W_g \right)_k - \sum_{j=1}^{N_b} \left(W_{app} \right)_j, \tag{6.17}$$

where

N_v = number of 3D volume elements
N_n = number of nodes
N_b = number of boundary surfaces on which forces act
E_i = energy of deformation associated with i^{th} volume element
$(W_g)_k$ = work done by gravity on the kth node
$(W_{app})_j$ = work done by surface forces on the jth surface node.

Let us describe each portion of 6.17 separately.

For a discrete model, nodes are distributed randomly within the region of space containing the material. This defines a set of internal nodes and a set of surface nodes. The space is then divided into nonoverlapping tetrahedra, each tetrahedron being defined by a node at each corner of the tetrahedron (see Fig. 6.2).

As a material is deformed, the nodes will move, distorting the tetrahedra and resulting in a change in the energy stored in each tetrahedron, E_i. We will describe the stored energy in terms of the stored energy per unit initial volume, ϵ_i, of each tetrahedron. That is,

$$E_i = \epsilon_i V_i, \tag{6.18}$$

where V_i is the initial volume of the ith tetrahedron. Then

$$\sum_{i=1}^{N_v} E_i = \sum_{i=1}^{N_v} \epsilon_i V_i \tag{6.19}$$

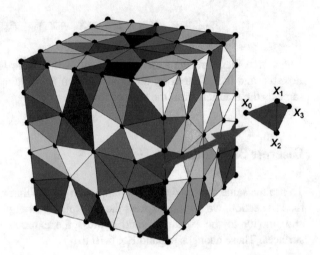

Fig. 6.2 A cube of material divided into tetrahedra. Each tetrahedron is defined by four corner nodes. In this figure, one tetrahedron in the material has been extracted so that the four corner nodes of the tetrahedron, X_0, X_1, X_2, and X_3, can be clearly labeled

is the total energy stored in the tetrahedra due to their deformation, and N_v is the number of tetrahedra simulated.

The gravitational pull on each tetrahedron can be expressed by choosing the direction of the force of gravity to be in the $-z$ direction. W_g becomes

$$W_g = \int_{Z_1}^{z_i} F_g \; dz = -m_k g(z_k - Z_k), \tag{6.20}$$

where

z_k is the current z component of the kth node
Z_k is the initial z component of the kth pode
m_k is the mass of the kth node
g is the acceleration due to gravity.

This gives the total gravitational energy of the material to be

$$-\sum_{k=1}^{N_n} \left(W_g\right)_k = +\sum_{k=1}^{N_n} m_k g(z_k - Z_k), \tag{6.21}$$

where N_n is the number of nodes in the simulation and the mass of each node, m_k, is found by taking the sum of one-fourth of each of the masses in all of the tetrahedra which contain the chosen node. As an example, consider the common case of the density of the material being uniform throughout the material. Suppose node 7 is found in three tetrahedra, numbered 1, 2, and 3. The mass which the tetrahedra contributes to node 7 is

$$m_7 = \rho_0 \frac{V_1 + V_2 + V_3}{4}, \tag{6.22}$$

where

$$\rho_0 = \frac{\text{total mass of the body}}{\text{total initial volume of the body}}. \tag{6.23}$$

For the case where node k is contained in N_k tetrahedrons, the total volume contributed to the node is

$$\overline{V}_k = \frac{\sum_{m=1}^{N_k} V_m}{4}. \tag{6.24}$$

where $\overline{V}_k$ is the volume associated with the kth node. It is assumed that mass is neither created nor destroyed as the deformation takes place, so that the mass associated with each node will not change during the deformation.

Thus

$$\sum_{k=1}^{N_n} (W_g)_k = -\sum_{k=1}^{N_n} \rho_0 g(z_k - Z_k)\overline{V}_k, \tag{6.25}$$

where the sum is over all the nodes in the material.

To calculate the work done by the forces on the surface of the material, $\sum_{j=1}^{N_b} W_{\text{app}}$, we need the force on each surface node. If the surface force, $\vec{F}_{\text{surf}}$, is applied uniformly over the surface, the force per unit initial surface area, $\vec{f}_{\text{app}}$, would be

$$\vec{f}_{\text{app}} = \frac{\vec{F}_{\text{surf}}}{A_0}, \tag{6.26}$$

where

A_0 = the initial total surface area $\vec{F}_{\text{surf}}$ is applied to.

To find the force on each surface node, the surface area associated with each node is needed. Each surface node has an area associated with it, A_j, defined as 1/3 the sum of the surface area of all the surface areas of the tetrahedra containing that node (e.g., Fig. 6.3).

The surface force on node j would then be, $\vec{f}_j$,

$$\vec{f}_j = \vec{f}_{\text{app}} A_j. \tag{6.27}$$

Following the same logic as with the spring model of the last section, the work done by all the nodes would then be

$$\sum_{j=1}^{N_b} (W_{\text{app}})_j = \sum_{j=1}^{N_b} \vec{f}_j \circ \left(\vec{x}_j - \vec{X}_j\right) A_j, \tag{6.28}$$

where

Fig. 6.3 Initial surface area, A_j, in gray that is associated with node shown as the gray sphere

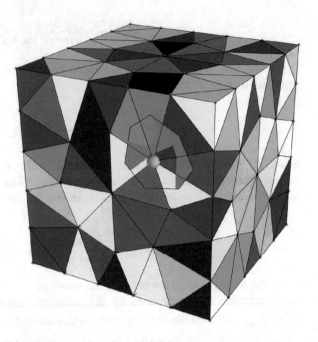

$\vec{x}_j$ = the vector denoting the current position of the jth node

$\vec{X}_j$ = the vector denoting the initial position of the jth node.

A_j = initial area associated with each node

N_b = number of boundary nodes forces are applied to.

Note that the force per unit initial area on node j, f_j, is the applied force "rescaled" by dividing by the initial surface area associated with the node. f_j is not the current pressure (or stress) on the node unless the initial surface area has not been changed by the deformation. During the deformation $\vec{F}_{\text{surf}}$ may change, but A_j remains constant since it is the initial area to which the force is applied. We have now

$$E_{\text{tot}} = \sum_{i=1}^{N_v} \epsilon_i V_i + \sum_{k=1}^{N_n} \rho_0 g(z_k - Z_k)\overline{V}_k$$
$$- \sum_{j=1}^{N_b} \vec{f}_{\text{app}} \circ \left(\vec{x}_j - \vec{X}_j\right)A_j. \qquad (6.29)$$

This energy is then minimized by calculating the change in E_{tot} as each node is displaced by a small amount and setting the change in E_{tot} equal to zero. This finds an extremum of E_{tot}, which we will assume is the minimum of E_{tot}. Note that when E_{tot} is minimized, $m_k g Z_k$ and $\vec{f}_{\text{app}} \circ \vec{X}_3$ in Eq. 6.29 are constants. Thus Eq. 6.29 can be simplified to

$$J = \sum_{i=1}^{N_v} \epsilon_i V_i + \sum_{k=1}^{N_n} \rho_0 g x_k \overline{V}_k - \sum_{j=1}^{N_b} \vec{f}_{app} \circ \vec{x}_j A_j. \qquad (6.30)$$

For simulating an isotropic body as described in Eq. 5.20,

$$\epsilon_i = f(\mathcal{I}_1, \mathcal{I}_2, \mathcal{I}_3). \qquad (6.31)$$

Minimizing J in Eq. 6.30 is the process we will use to numerically solve the problems in the next two chapters. To minimize J set

$$\frac{\partial J}{\partial x_i} = 0 \qquad (6.32)$$

for all of the components of all of the nodes not fixed by Dirichlet (fixed) boundary conditions.

Continuous 3D Model

It is customary to write discrete equations in a continuous form. To define the continuous equations, we need to take the limit as the number of nodes and the number of tetrahedra approach infinity as the volume associated with each node and the volume of each tetrahedra approach zero. At the same time, the area associated with each node on the surface will also approach zero. Equation 6.30 would then be written as

$$J = \lim_{\substack{N_v \to \infty \\ V_i \to 0}} \sum_{i=1}^{N_v} \epsilon_i V_i + \lim_{\substack{N_n \to \infty \\ V_k \to 0}} \sum_{k=1}^{N_n} \rho_0 g z_k \overline{V}_k$$

$$- \lim_{\substack{N_b \to \infty \\ A_j \to 0}} \sum_{j=1}^{N_b} \vec{f}_{app} \circ \vec{x}_j A_j \qquad (6.33)$$

or

$$J = \int_{volume} \epsilon \, dV + \int_{volume} \epsilon_g \, dV - \int_{surface} \vec{f}_{app} \circ \vec{x}_b dA, \qquad (6.34)$$

where

$$\epsilon = f\left(\frac{\partial x_i}{\partial X_j}\right) \tag{6.35}$$

$$\epsilon_g = \rho_0 g z \tag{6.36}$$

and

$\vec{x}_b$ = vector defining the location of points on the boundary surface.

The solution is found by minimizing J, which is usually written in shorthand as

$$\delta J = 0. \tag{6.37}$$

To carry out this minimization for a continuous equation requires the calculus of variations. Fortunately for numerical simulations, we only need to minimize the discrete energy equation, Eq. 6.30.

To apply Eq. 6.32 to Eq. 6.30 in the discrete equation, we need to write $\frac{\partial x_i}{\partial X_j}$ in terms of x_i and X_j so that we can carry out the derivative given in Eq. 6.32. This process will be described in Chap. 7.

Problems

Problem 1 Consider the configuration of nodes and springs shown in Fig. 6.1. Let the rest length of the two springs be 1 m. The springs themselves are identical and massless. Each node has a mass of 2 kg. Assume the spring constant of each spring is 100 N/m. Hang the mass spring system vertically from node 1, where node 1 is at (0,0,0), so that gravity tends to extend the springs. Use the summation of forces on each node to solve for the final positions of the spring masses and the force from the ceiling on node 1.

Problem 2 Repeat problem 1, but this time use the minimization of Eq. 6.10 to find the final positions of the nodes and the force from the ceiling. (Hint: Consider the gravity force as the applied force which does work on the nodes.)

Problem 3 Repeat problem 1, but this time use the equation of motion, Eq. 4.45, to find the final positions of the nodes and the force from the ceiling. (Hint: The ϵ for a single spring is found in Eq. 4.23, and the acceleration of each node is zero.)

Problem 4 Given a constant force of (0,3 N, −2 N) applied to the top of a 12 g cube 1 cm × 1 cm × 1 cm, calculate the work done by this force as the cube is compressed only in the z direction to become 1 cm × 1 cm × 0.75 cm.

Problem 5 Given the cube and deformation in problem 4, calculate (a) the mass per unit original volume,ρ_0, and (b) the true density of the cube after the deformation, ρ.

Problem 6 Consider a 5 g unit cube represented by nodes at the eight corners of the cube (p_1 through p_8) and one node inside the cube at $p_9 = (0.5,0.5,0.6)$. The cube can be divided into the following 12 tetrahedra:

{{p9,p5,p1,p3}, {p6,p2,p1,p9}, {p8,p5,p9,p3}, {p1,p7,p9,p3}, {p6,p8,p2,p9}, {p1, p2,p5,p9}, {p2,p8,p5,p9}, {p9,p7,p1,p4}, {p9,p1,p6,p4}, {p9,p7,p8,p3}, {p8, p9,p6,p4}, {p8,p7,p9,p4}}.

Calculate the volume associated with p4, where p4 $= (0.,0.,1.)$.

Problem 7 Given the cube in problem 6, find the surface area associated with node 4.

Problem 8 Consider the deformation described in Fig. 4.4 by three moveable nodes. If the applied force was 7 and the springs could be described with $F = -k\chi$, what is k?

Problem 9 Show that if z is chosen in the same direction as the force of gravity (i.e., down, $-z$), then the potential energy due to gravity which contributes to J in Eq. 6.30 is $-$ m g z instead of $+$ m g z.

Problem 10 Find the contribution to J of the applied force, $\vec{F}_{applied}$, with magnitude 2 N applied to an area of $1/2m^2$ in the following cases

(a) $\vec{F}_{applied}$ up, +z up

(b) $\vec{F}_{applied}$ up, +z down

(c) $\vec{F}_{applied}$ down, +z up

(d) $\vec{F}_{applied}$ rotated by 20^0 from the vertical with

$$\mathscr{R} = \begin{pmatrix} \cos\left(20^0\right) & 0 & \sin\left(20^0\right) \\ 0 & 1 & 0 \\ -\sin\left(20^0\right) & 0 & \cos\left(20^0\right) \end{pmatrix},$$

i.e., $F_{appied} = (2N)\mathscr{R}\,\hat{k}$, and +z up.

Chapter 7
Simulations

Introduction

This chapter describes the details of how to find the minimum of Eq. 6.30 using Mathematica. The basic concept is to describe the material in terms of a set of nodes (e.g., Fig. 6.2), express the $\frac{\partial x_i}{\partial X_j}$ in terms of these nodes, and then take the derivative of J with respect to every node's x, y, and z components. The result of each derivative is then set to zero to minimize J with respect to each node's position. The minimum is found by changing the values of all the node components until J does not change value within some tolerance. This can be a daunting task since simulations may have hundreds of nodes. However, modern computer software makes this possible. For example, Mathematica has a command called **FindMinimum**. Input the function you wish to minimize (for us it is J) and a list of the variables (for us it is all the node's components which are not fixed), and **FindMinimum** minimizes the input function! There are of course a few details, which we will need to discuss.

In the following, specific references are made to the Mathematica code used to produce example results. These are in bold print, like the word **FindMinimum** in the previous paragraph. The actions of the Mathematica commands are described as we encounter them. Parameter lists and defined subroutines are also bold. Mathematica's built-in subroutines always begin with a capital letter. User-defined subroutines always begin with a small letter.

The necessary steps are as follows:

1. Input variables to define the problem, the geometry, and the boundary conditions.
2. Nodes are distributed within the geometry chosen, and nonoverlapping tetrahedra are constructed to cover the entire material of interest.

Supplementary Information The online version contains supplementary material available at [https://doi.org/10.1007/978-3-031-09157-5_7].

3. The position boundary conditions (Dirichlet) define the final location of chosen components of the boundary nodes.
4. The force boundary conditions (Neumann) define the force on chosen boundary nodes.
5. Initial volumes of each tetrahedra are calculated.
6. The deformation energy per unit initial volume, ϵ, is input as a function of I_i or $\frac{\partial x_i}{\partial X_j}$. This is converted into a function of node locations. The energy contribution from each tetrahedra is calculated as the product of the energy per unit initial volume times the initial volume for each tetrahedra. The sum of the energy contribution from all the tetrahedra is the total deformation energy.
7. A list is made of the coordinates of those nodes which are free to move.
8. The energy associated with gravity is calculated for each node.
9. The energy from the force boundary conditions and gravity is added to the deformation energy to define J. Then all the node component values are varied until J is a minimum.
10. A plot is made of results from the final positions of the nodes

This chapter describes each of these steps and the details of simulating a nearly incompressible materials. There will be variations of each of these steps depending upon the exact problem solved, but these steps form a bases of all the simulations. In Chap. 8, variations of this procedure will be discussed in each example.

User Input

The user input usually consists of the following:

runName = name of the run which will be saved with output.
ρg = material density (mass per initial volume) times g (the acceleration due to gravity).
i123eq = ϵ in terms of I_i's or $\frac{\partial x_i}{\partial X_j}$ values. Note that **i123eq** and **ρg** must have compatible units.
maxforce = the applied surface force after all steps have been executed (if cylinder or cuboid).
area = relative number to define the number of nodes in the object.
nsteps = number of steps to reach the final deformation (remember forces and displacements must be turned on in steps to find the physical solution to define a continuous path between the initial and final configurations).
frac = fractional thickness of cylinder or cuboid after all steps have been executed.

The object to be deformed is defined by calling one of the following routines:
2D

rectangleRegularPts = builds a rectangle of regularly spaced nodes
rectangleRandomPts = builds a rectangle of randomly spaced nodes
diskRegularPts = builds a disk of regularly spaced nodes
diskRandomPts = builds a disk of randomly spaced nodes

3D

cuboidRegularPts = builds a cube of regularly spaced nodes
cuboidRandomPts = builds a cube of randomly spaced nodes
cylinderRegularPts = builds a cylinder of regularly spaced nodes
cylinderRandomPts = builds a cylinder of randomly spaced nodes
sphereRandomPts = builds a sphere of randomly spaced nodes

The boundary conditions can also be input using the following arrays:

comploc = final values of the fixed components of the constrained nodes
force = values of the maximum force per unit initial area that are applied at each node or over a surface of the object.

Alternatively, for deformations of cylinders and cuboids, either **maxforce** or **frac** is used to define the maximum force on the top of the object or the final fractional thickness of the material after the deformation is complete. The input of **maxforce** and **frac** is exclusive. If **maxforce** is nonzero, **frac** should be zero, and if **frac** is nonzero, **maxforce** should be zero, since these correspond to either Neumann or Dirichlet boundary conditions.

Define Nodes and Their Connectivity

Nodes are distributed within the volume and a connectivity is defined. This process makes heavy use of Mathematica's mesh building software. For example, a call may be made to populate a cylinder (**Cylinder**) or a sphere (**Ball**) with nodes on the surface using **surfacemeasure** to define the density of these nodes. Next **TriangulateMesh** is called to define nodes and connectivity within the material. Mathematica creates a Delaunay-style mesh, connecting the nodes so that each set of four nodes define a unique nonoverlapping tetrahedra within the material. These tetrahedra are created so that they are not too elongated and the distribution of the size of the tetrahedra is relatively narrow. These two conditions are important to help prevent inversion of any tetrahedra during simulations. (Remember there are multiple solutions to the nonlinear differential equations which define deformation.)

Nodes are defined by first defining surface nodes. Then the internal nodes are added. A mesh is created with all the nodes. A list of nodes is returned, **pts**, along with a list of tetrahedra containing four nodes for each tetrahedra (**poly**) and the number of boundary nodes (**nbpts**). The surface nodes are listed first within the node list so that the first **nbpts** in **pts** are the surface nodes. The surface nodes are listed this way because boundary conditions are applied only to these nodes. The form of this is {**pts,polys,nbts**}. A plot is then made to verify that the nodes have been created as intended. The plot is also useful for finding the node numbers where boundary conditions should be applied. An example plot for a cylinder is shown in Fig. 7.1.

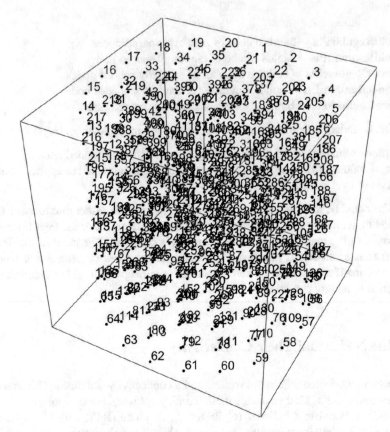

Fig. 7.1 An example of a cylinder with nodes and node numbers displayed

Position Boundary Conditions

There are two types of boundary conditions. One type of boundary condition sets the
final location of a node. The other sets the final force on a node. These are entered as
a list in **comploc** for fixed nodes and in the **force** array for forces. Both lists consist
of a list of nodes followed by the final location or maximum force on each node. For
example, **comploc** = {{2,{1,1,1}},{5,{1,2,3}},...} tells the simulation to end the
simulation with node 2 at location {1,1,1} and node 5 at location {1,2,3}. **force**=
{{7,{1,.1,.6}} tells the simulation to end the simulation with node 7 having a force
on it of {1,.1,.6}. In both cases, the units are defined by the dimensions of the created
object for node locations and by the dimensions of ϵ. If not all the components of a
node are to be fixed, the input for those components that are not to be fixed is entered
as "free". For example, if only the x component of node 2 is fixed and the y and z
components are free to change, **comploc** is defined as **comploc** = {{2,{1,free,free}},
{5,{1,2,3}},...}. Forces are not allowed this option, and the force in x,y, and z

directions must be set for each node if a force constraint is applied to that node. If neither force nor position constraint is added for a particular boundary node, the boundary condition of that node is a force of zero.

Two sets of variables are used within the simulation to define node locations. {**X [i],Y[i],Z[i]**} define the initial location of the nodes and {**x[i],y[i],z[i]**} define the final location. All of the **X[i]** values were set in step 2. The final node values which are constrained are defined using the following pattern:

$$\mathbf{x[i] = X[i] + step/nsteps\ (finalpt[i] - X[i])}, \tag{7.1}$$

where i is chosen for each node value in **comploc** and **finalpt** is the value that follows the node value in **comploc**. Equation 7.1 slowly changes the position of node i, $X[i]$, to the final location of the node at **finalpt[i]**. As step varies from 1 to **nsteps**, Eq. 7.1 changes the positions of the nodes in equal steps from their initial positions **X[i]** to their final positions. The maximum number of steps is **nsteps**.

maxforce is used to define the maximum force on the top surface in a simulation of a cuboid or cylinder. It is converted into **force** (the force on each node) by multiplying the **maxforce** times the area associated with each node divided by the total area of the top surface. The area associated with each node is found by summing the area of all the polygons containing a node. This sum is divided by 3 to assign 1/3 of the area of each polygon containing the node to that node. This approach distributes the **maxforce** so that the force per unit initial area is the same over the entire top surface.

Force Boundary Conditions

Each force in the force list, force, is dotted with the node location that the force is applied to. This product is multiplied by step/nsteps to slowly "turn on" these boundary forces using the following pattern:

$$\mathbf{bcs = (step/nsteps) * force[[i, 2]].\{x[fpt], y[fpt], z[fpt]\}}. \tag{7.2}$$

The contributions of the boundary forces are summed and stored in a variable called bcs. A plot is then made of the boundary conditions superimposed on the initial grid. An example is shown in Fig. 7.2.

Calculate Initial Volumes

The initial volume of each tetrahedron is calculated, V_k. **polys** contains the node locations which define each tetrahedra. So, for example, if **polys** = {{1,2,3,20}, {2,4,7,12}}, two tetrahedra are defined, the first with the four nodes located at 1, 2,

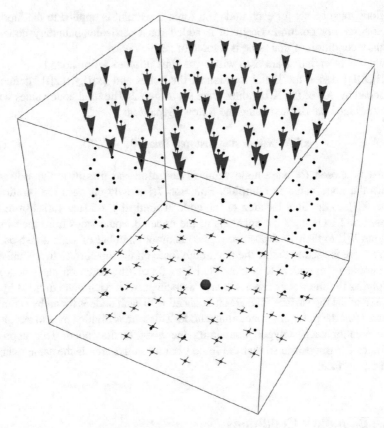

Fig. 7.2 The boundary conditions are displayed. Black nodes are free to move in any direction. Nodes with short red lines are only allowed to move in the direction of the lines. The node with the red ball is not allowed to move in any direction. Forces are shown as blue arrows. Node movements are shown as red arrows, but none are displayed here. This display is made to help the user check if the boundary conditions were implemented as intended

3, and 20 positions in **pts** and the second with nodes located at positions 2, 4, 7, and 12. Knowing the coordinates of four nodes, in the first set of points, pt1, pt2, pt3, and pt20, allows the definition of three vectors (see Fig. 7.3),

$$
\begin{aligned}
\vec{v}_{12} &= \text{pt2} - \text{pt1} \\
\vec{v}_{13} &= \text{pt3} - \text{pt1} \\
\vec{v}_{14} &= \text{pt20} - \text{pt1}.
\end{aligned} \tag{7.3}
$$

The initial volume of each tetrahedron is calculated as

$$
V_k = \left| \left(\vec{v}_{12} \times \vec{v}_{13} \right) \cdot \vec{v}_{14} \right| / 6. \tag{7.4}
$$

Fig. 7.3 Tetrahedron
defined by four nodes with
vectors used to calculate the
volume of the tetrahedron

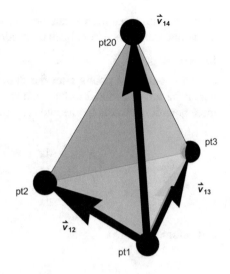

Calculate Deformation Energy per Unit Initial Volume

The user input for deformation energy per unit initial volume, ϵ, is given as a function of either the invariants, I_i, or the derivatives, $\frac{\partial x_i}{\partial X_j}$. If the function is in terms of I_i, I_i must be converted into a function of $\frac{\partial x_i}{\partial X_j}$ using Eq. 5.21. Then $\frac{\partial x_i}{\partial X_j}$ is converted into node positions. The components of the final position, (x_i for $i = 1,2,3$), of any point in the material are a function of the components of the initial position, (X_i for $i = 1,2,3$), i.e.,

$$x_i = f_i(X_1, X_2, X_3) \tag{7.5}$$

The differential of the ith component is

$$dx_i = \frac{\partial x_i}{\partial X_j} dX_j \quad (\text{summed over } j = 1, 2, 3). \tag{7.6}$$

In full form this is

$$
\begin{aligned}
dx_1 &= \frac{\partial x_1}{\partial X_1} dX_1 + \frac{\partial x_1}{\partial X_2} dX_2 + \frac{\partial x_1}{\partial X_3} dX_3 \\
dx_2 &= \frac{\partial x_2}{\partial X_1} dX_1 + \frac{\partial x_2}{\partial X_2} dX_2 + \frac{\partial x_2}{\partial X_3} dX_3 \\
dx_3 &= \frac{\partial x_3}{\partial X_1} dX_1 + \frac{\partial x_3}{\partial X_2} dX_2 + \frac{\partial x_3}{\partial X_3} dX_3
\end{aligned} \tag{7.7}
$$

If we assume that each tetrahedron in calculation is uniform, there is exactly one $\mathcal{F}_{ij}$ within each tetrahedron. Each tetrahedron is defined by initially by the location of four corner points, $\vec{X}^{(1)}, \vec{X}^{(2)}, \vec{X}^{(3)}$, and $\vec{X}^{(4)}$ before the deformation and the location of these same four points after the deformation, $\vec{x}^{(1)}, \vec{x}^{(2)}, \vec{x}^{(3)}$, and $\vec{x}^{(4)}$. The tetrahedra are considered to be small so that dx_i can be approximated by defining three bounding vectors before and after the deformation:

$$
\begin{aligned}
\vec{dx}^{(1)} &= \vec{x}^{(2)} - \vec{x}^{(1)} \\
\vec{dx}^{(2)} &= \vec{x}^{(3)} - \vec{x}^{(1)} \\
\vec{dx}^{(3)} &= \vec{x}^{(4)} - \vec{x}^{(1)}
\end{aligned}
\tag{7.8}
$$

and before

$$
\begin{aligned}
\vec{dX}^{(1)} &= \vec{X}^{(2)} - \vec{X}^{(1)} \\
\vec{dX}^{(2)} &= \vec{X}^{(3)} - \vec{X}^{(1)} \\
\vec{dX}^{(3)} &= \vec{X}^{(4)} - \vec{X}^{(1)}
\end{aligned}
\tag{7.9}
$$

Each of these vector equations has three components, yielding nine equations when used in the above set of three equations:

$$
\begin{aligned}
dx_1^{(1)} &= \frac{\partial x_1}{\partial X_1} dX_1^{(1)} + \frac{\partial x_1}{\partial X_2} dX_2^{(1)} + \frac{\partial x_1}{\partial X_3} dX_3^{(1)} \\
dx_2^{(1)} &= \frac{\partial x_2}{\partial X_1} dX_1^{(1)} + \frac{\partial x_2}{\partial X_2} dX_2^{(1)} + \frac{\partial x_3}{\partial X_3} dX_3^{(1)} \\
dx_3^{(1)} &= \frac{\partial x_3}{\partial X_1} dX_1^{(1)} + \frac{\partial x_2}{\partial X_2} dX_2^{(1)} + \frac{\partial x_3}{\partial X_3} dX_3^{(1)} \\
dx_1^{(2)} &= \frac{\partial x_1}{\partial X_1} dX_1^{(2)} + \frac{\partial x_1}{\partial X_2} dX_2^{(2)} + \frac{\partial x_1}{\partial X_3} dX_3^{(2)} \\
dx_2^{(2)} &= \frac{\partial x_2}{\partial X_1} dX_1^{(2)} + \frac{\partial x_2}{\partial X_2} dX_2^{(2)} + \frac{\partial x_2}{\partial X_3} dX_3^{(2)} \\
dx_3^{(2)} &= \frac{\partial x_3}{\partial X_1} dX_1^{(2)} + \frac{\partial x_3}{\partial X_2} dX_2^{(2)} + \frac{\partial x_3}{\partial X_3} dX_3^{(2)} \\
dx_1^{(3)} &= \frac{\partial x_1}{\partial X_1} dX_1^{(3)} + \frac{\partial x_1}{\partial X_2} dX_2^{(3)} + \frac{\partial x_1}{\partial X_3} dX_3^{(3)} \\
dx_2^{(3)} &= \frac{\partial x_2}{\partial X_1} dX_1^{(3)} + \frac{\partial x_2}{\partial X_2} dX_2^{(3)} + \frac{\partial x_2}{\partial X_3} dX_3^{(3)} \\
dx_3^{(3)} &= \frac{\partial x_3}{\partial X_1} dX_1^{(3)} + \frac{\partial x_3}{\partial X_2} dX_2^{(3)} + \frac{\partial x_3}{\partial X_3} dX_3^{(3)}
\end{aligned}
\tag{7.10}
$$

All of the $dx_i^{(k)}$ and $dX_i^{(k)}$ values are calculated from the initial and final positions of each of the four corners of each tetrahedron: therefore in the nine equations above, there are exactly nine unknowns ($\frac{\partial x_i}{\partial X_j}$). Independence is guaranteed as long as the tetrahedron has nonzero volume. The final node positions are not known, but Equation 7.10 is solved and the values of $\frac{\partial x_i}{\partial X_j}$ are stored as functions of the final node positions for each tetrahedron.

The deformation energy per unit initial volume for each node, ϵ_k, is now a function of the unknown node positions. The total deformation energy, etot, is the sum of the product of ϵ_k for each node times the initial volume of each tetrahedron.

Define Variable List

Because of the fixed boundary nodes, not all final node positions are unknown. Only the unknowns can be sent to **FindMinimum**. A list of unknown variables is created by first creating a list of all the nodes and all of their coordinates:

$$\textbf{varsi} = \text{table of}\,\{x[i], y[i], z[i]\}\text{values} \qquad (7.11)$$

with i varying from 1 to the total number of nodes in the simulation. Then the components of the nodes in **comploc** which are not "free" are removed from **varsi**. What remains in varsi are the components of the nodes that are free to move.

Gravity

The gravitational energy associated with each node is calculated as follows.

Each node is assigned a volume, **V[i]**, as 1/4 the sum of the volumes of all the tetrahedra having that node as a corner. The force of gravity on that node is calculated as $\rho\textbf{g}\,\textbf{V[i]}$. The total gravity contribution to J is the sum of all these terms.

$$\textbf{gravity} = \sum\nolimits_{i=1}^{N}\rho\textbf{g}\textbf{V}[i] \qquad (7.12)$$

Notice that the volume which multiplies ϵ is the volume of each tetrahedron, whereas the volume which multiples $\rho\textbf{g}$ is the volume associated with each node.

Go

J is minimized by calling FindMinimum in Mathematica with

$$J = (\textbf{etot} + \textbf{gravity}) - \textbf{bcs} \qquad\qquad (7.13)$$

using the set of variables, **varsi**.

The boundary conditions are applied in small steps, and a conjugate gradient method is used to find the minimum. J is a nonlinear function with multiple minimum values. The steps in Eqs. 7.1 and 7.2 ensure that there is a continuous path of change from the initial to the final state. A maximum of 10,000 iterations is used for each step. Often this number of iterations is not enough to reach convergence within machine tolerances. When this happens, the program returns the following warning:

```
··· FindMinimum:
 "The line search decreased the step size to within the tolerance specified by AccuracyGoal and
  PrecisionGoal but was unable to find a sufficient decrease in the function. You may need more
  than \!\(\*RowBox[{\"MachinePrecision\"}]\) digits of working precision to meet these tolerances."
```

Even though machine tolerance may not have been reached, most of the time the answer is still accurate enough for applications. A better measure of the accuracy (or at least the repeatability) of our result is to run test cases increasing the number of steps and nodes until the results are the same within the desired tolerance. The number of iterations and the machine precision of the numerical calculation can be increased with options in **FindMinimum**, should this be necessary for a specific problem.

Plots

Plots are made of the results of the simulation. These plots have been copied directly into this text for each example simulation. Depending upon the problem, plots may be made of final locations of the nodes or other calculated results. Chapter 13 provides some additional calculations that might be useful depending upon the problem to be solved.

Simulating Incompressible Materials

To model incompressible (or nearly incompressible) materials, the term $\gamma (I_3 - 1)^2$ is added to ϵ with γ large compared to the other terms in ϵ. The larger the value of γ relative to the other terms, the closer to incompressibility the simulation will produce; however, too large a value of γ will make the equations too "stiff" to solve by conjugate gradient techniques. Thus, the γ value in simulations should be limited to a range of 100–10000 times the other multiples of the I_i terms in ϵ.

Although it is possible to force incompressibility in Mathematica by setting $I_3 = 1$ as a constraint in **FindMinimum**, in that case, Mathematica no longer allows the use of the conjugate gradient solution technique. Other **FindMinimum** algorithms tend to find solutions not near the starting values. Nonlinear functions can have many solutions for a single set of force boundary conditions. To simulate material deformations, it is necessary to take small steps, finding the solution in each step with the least changes in node locations. In each step the conjugate gradient technique is used to find a solution "near" the solution found in the last step.

Scaling in Simulations

The simulation of incompressible materials can involve many orders of magnitude. For example, the simulation of Dragon Skin material involves

$$\epsilon = b\,\mathcal{I}_1 + \gamma(\mathcal{I}_3 - 1)^2, \tag{7.14}$$

with $b = 3.4^*10^4$ N/m^2 and $\gamma = 3.4^*10^6$ N/m^2

(see Eqs. 10.30, 10.31 and 10.32) and material sizes on the order of 10^{-2} m. This means that there is as much as eight orders of magnitude difference in the input values using the units of Newtons and meters. Mathematica seems to have no difficulty in working with these disparate values, but for some computer software, this large a difference can create round-off problems. These can be avoided by using dimensionless units.

There are only two fundamental units in a quasi-static deformation – force and distance. By making these dimensionless or equivalently, using computer simulation dimensions, large discrepancies in the numerical values can be avoided during the calculations. Once the computations are completed, the answers can be converted back from computer simulation units to the normal dimensional units. An example is perhaps the easiest way to see how this works.

In the simulation of a Dragon Skin cube in Fig. 8.14, Eq. 7.14 defines the energy per unit initial volume, and the size of the cube simulated is 0.0124 m. Instead of using Newtons and meters, computer units, cF, for computer force units and cD for computer distance units can be defined such that 1 cD $=$ 0.0124 m and 1 cF $= 3.1^*10^4$ N/m^2*(0.0124 m)$^2 = 5.22784$ N, so that in computer units, the cube size is now 1 cD and b $=$ 1 cF/cD^2 and $\gamma =$ 100 cF/cD^2. The simulation can now be carried out in computer units. Note that now there is only two orders of magnitude difference in the user numerical inputs in the computer unit simulations. This should pose no problems for any computer code.

Once the final answer is found in computer units, the numerical values of the simulation can be returned to real-world units by multiplying all distance in cD by 0.0124 m/cD and all force in cF by 5.2278 N/cF. The process of converting to dimensionless units can be used for any size and type of material.

Problems

Problem 1 Given that the volume of a tetrahedron is 1/3 * (area of the base)* (height) and that the area of a triangle is 1/2 the area of a rectangle, show if $\vec{v}_{12}$, $\vec{v}_{12}$, and $\vec{v}_{12}$ are vectors with a common origin that define the sides of a tetrahedron, $(\vec{v}_{12} \times \vec{v}_{13})$. $\vec{v}_{12}$ /6 gives the volume of a tetrahedron.

Problem 2 Solve for $\frac{\partial x_i}{\partial X_j}$ using

$$x_i = \frac{\partial x_i}{\partial X_j}X_j + b_i$$

given the initial three points:

$$\text{pt}_1 = (1, 2), \text{pt}_2 = (3, 1), \text{pt}_3 = (4, 4),$$

and the final three points

$$\text{pt}_1 = (1.7, 2), \text{pt}_2 = (4.1, 2), \text{pt}_3 = (4.6, 4.8)$$

respectively.

Problem 3 Find I_1 given the mapping

$$x = 2X + 3Y - 1Z + 4$$
$$y = 1X + 2Y + 2.5X - 3.$$
$$z = .5X - 1Y + 3X - 3.$$

Problem 4 Calculate energy in terms of $\frac{dx_i}{dX_j}$ for $\epsilon = I_3$.

Problem 5 Find the initial volume of tetrahedron given four corner points (0, 0, 0), (1, 0, 0), (0, 1, 0), and (0, 0, 1).

For problems 6 through 10, consider a unit square defined by five points and four triangles as shown below, with $p_1 = (0,0)$, $p_2 = (1,0)$, $p_3 = (0,1)$, $p_4 = (1,1)$, and $p_5 = (0.5,0.5)$.

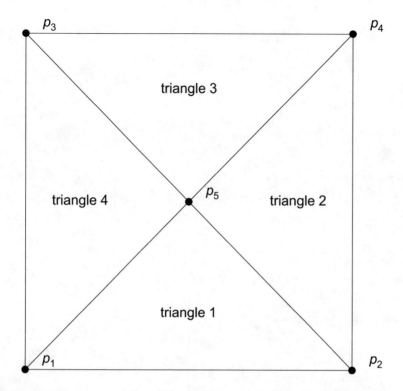

Problem 6 Create the list {points, polygons, number of boundary points} for the figure.

Problem 7 Create the comploc listing if p1 and p2 are fixed in the x direction and p1 moves in the y direction +0.1.

Problem 8 Assume p_1 has a final position after 20 steps of $(0, 0.1)$. Find the location of the point p_1 after four steps.

Problem 9 Assume a force of 3 is to be applied in the -y direction to the top of the rectangle. Write the force constraint, bcs, for p_3 after step 3 of 20 equal steps.

Problem 10 Assume the deformation energy per unit initial volume of the square is defined by

$$\epsilon = 1/2(200) \left(\frac{\partial x_1}{\partial x_1} + \frac{\partial x_2}{\partial x_2} \right)^2.$$

Express the energy of triangle 1 as a function of the coordinates of each point.

Chapter 8
Quasi-static Simulation Examples

Introduction

This chapter contains seven example simulations for quasi-static processes. The examples are designed to show how increasing complexity can be programmed. The examples are as follows:

1. Single tetrahedron study (to quickly study the effect of different ϵ)
2. Cylindrical compression test (to compare the results of the single tetrahedron study to many tetrahedra)
3. Real compression and extension
4. Matching a physical deformation
5. Bending paper (2D/3D fit. Add energy of curvature)
6. Extrusion study (2D only. Model plastic behavior)
7. Water drop (2D and 3D. Add surface tension)

Each example is designed to show a variation in the solution of the basic equations, i.e., Eqs. 6.30 and 6.32. The codes are not included in the printed text but can be downloaded from the website. The purpose of the codes is not to provide a black box to be applied to application problems but instead are to show all the steps of each case. It is difficult to include every step in a discussion (that gets a bit boring); therefore, the complete information in the codes is provided as the code itself. The text gives a short summary of each new feature added to the code. The codes themselves have not been tested, debugged, or optimized to the level of commercial use but instead are pedagogical in nature.

The first example solves the force-displacement relationship of a single grid for a quasi-static process. The results are compared to analytic solutions of Eq. 6.32. The purpose of these simulations is to test the code to ensure that the code is solving

Supplementary Information The online version contains supplementary material available at [https://doi.org/10.1007/978-3-031-09157-5_8].

Eq. 6.32 accurately. A single grid model is not intended to be a physical model but instead allows the exploration of many values of the parameters in ϵ. Several different energy functions are simulated. Linear elastic behavior, strain hardening, and strain softening results are shown. The code from this simulation can be easily expanded to other energy functions.

The second example shows that the results of many-grid simulations are the same as the results of the single grid study. A cylindrical geometry is used for this study since cylindrical shaped pieces of material are often used in the laboratory to determine material properties. The boundary conditions allow the material to expand in the lateral direction without restriction. Unfortunately, in real experiments, the friction at the top and bottom of the cylinder restricts the lateral expansion of the material. This restriction is added in the third example.

The third example explores a more realistic deformation. Friction at the top of the cylinder in a real compression experiment results in a different force-displacement curve than the ideal simulated in the second example. In this example, both a cubic and cylindrical geometry are simulated so that a visual comparison can be made showing that the deformations are similar to physical examples.

The fourth example reproduces the deformation of Play-Doh shown in Fig. 2.1. This is done by assuming an elastic material, but the simulation could be easily changed to a plastic material (which Play-Doh actually is) using the re-griding technique described in the sixth simulation example in this chapter.

The fifth example is the simulation of a flat object that can bend in three dimensions. Sometimes this is called a 2 1/2 dimensional study. For this type of study, all points lie in a surface that is embedded in three dimensions. The surface can be deformed in 3D, but always remains only a 2D surface. This simulation requires adding a bending energy based on the curvature of the surface. In this example, the parameters in the bending energy are varied to fit the results of bending a piece of paper.

The sixth example is a 2D study of a 3D process. In this case a 2D study is made of extrusion. Two-dimensional studies very seldom correspond exactly to physical processes. Even so, 2D studies are often useful because many more grids can be solved with less computational power in 2D than in 3D. An exploration of many different scenarios in 2D can provide insight into real 3D problems.

The seventh and last example introduces surface tension into the solving of Eq. 6.32. It is done in both 2D and 3D. In this case, a simulation is made of a drop of water forming from a slowly leaking faucet. It simulates a material (water) without shear strength in a quasi-static process. For this example, surface tension is added to the simulation.

Single Grid Study

It has been said of computer code that there is always at least one more bug in the code. In other words, numerical simulations can easily produce erroneous results and must be tested as much as possible against analytic solutions and experiments. In this

Fig. 8.1 The parallelepiped
before and after a
deformation of the four-
node model

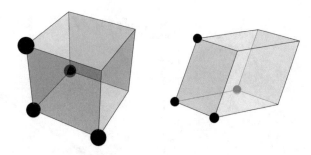

section a study is made of a single tetrahedron, both using the numerical computer code and an analytic solution. This section is strictly a mathematical exercise and has no "real-world" application other than testing if the algorithms used to solve Eq. 6.32 are correct. In later example simulations, we will increase the number of grids and the complexity of the geometry. This will allow comparison between simulations and experimental results.

In this simulation the force on a single grid is increased step by step to a maximum applied force. The displacement of each step versus the applied force at each step is plotted. Since there is only one grid, this exercise can be repeated for many different energy functions very quickly. An affine mapping (Eq. 7.6) is assumed so that the positions of four nodes are all that are required to completely define the deformation. The positions of these four nodes allow the calculation of all nine elements in the deformation gradient matrix, $\frac{\partial x_i}{\partial X_j}$. One node will be chosen at $(0, 0, 0)$ and remain fixed. The other three nodes will be chosen at $(1, 0, 0)$, $(0, 1, 0)$, and $(0, 0, 1)$. To remove pure rotations, the node along the x-axis will be allowed to move only in the x direction, and the node along the y-axis will be allowed to move only in the x-y plane. These restrictions will ensure that the cube does not rotate or translate. The remaining movements of these nodes can describe any deformation of the grid (Fig. 8.1).

In carrying out the simulation, there will be no "internal" nodes because the locations of all interior points are uniquely defined by the positions of the four nodes. Gravity will be zero. A force will be applied only along the z-axis against the z face of the cube. The applied force is chosen to be positive if it is compressive (i.e., in the − z direction on the top face and + z direction on the bottom face) and negative if it is extensional. The z component of the node along the z-axis is subtracted from one (i.e., $1-z$). This will display compression along the positive abscissa (x positive) and extension in the negative abscissa direction (x negative). The force is plotted as the ordinate (y-axis). An example plot is shown in Fig. 8.2. In this figure and in the following similar figures, the energy per unit initial volume, $\in$, is expressed in terms of the invariants, I_1, I_2, and I_3 corresponding to I_1, I_2, and I_3.

Notice in Fig. 8.2 that there is a slight upward curvature of the force-compression and force-extension plot. A more linear plot is shown in Fig. 8.3.

For students familiar with linear elasticity (Chap. 15), a plot of the linear force-displacement is made using Young's modulus from Eqs. 15.23, 15.24, 15.34, and

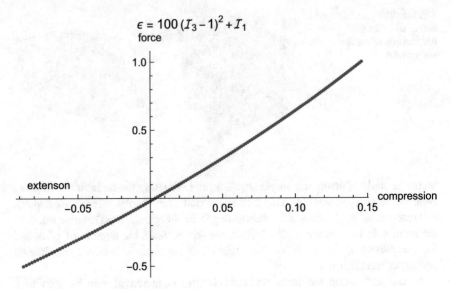

Fig. 8.2 Force versus the amount of compression (or extension) in the z direction using the energy function, ϵ, shown (I_i in the Figures corresponds to I in the text)

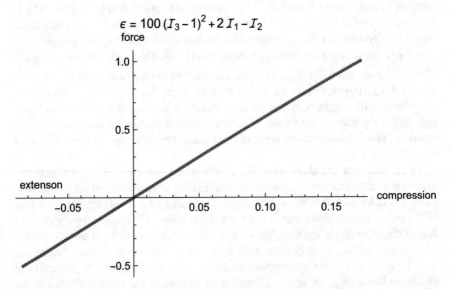

Fig. 8.3 Force versus the amount of compression (or extension) in the z direction using the energy function, ϵ, shown

15.35. The slope of the calculated curve from the four-node model should match the slope of the Young's modulus curve at the origin. Figure 8.4 shows that this is the case. The simulated case differs from the linear Young's modulus curve for $x > 0.15$ because Young's modulus is only defined for small deformations.

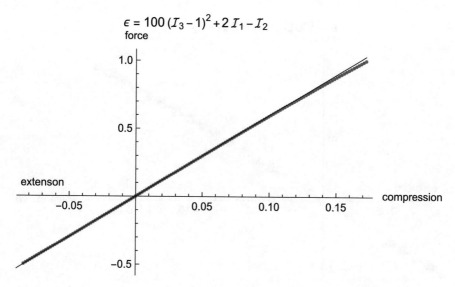

$$\epsilon = 100\,(I_3 - 1)^2 + 2\,I_1 - I_2$$

Fig. 8.4 Young's modulus calculated from the energy per unit volume, ϵ, and plotted in blue versus the numerically simulated force versus compression (or extension) in red

For students not familiar with linear stress and strain, it is sufficient to note that the relationship between force and compression in Fig. 8.3 is effectively linear for $x < 0.15$.

One more numerical test is carried out. Since no shear has been applied to the unit cube, the values of $\frac{\partial x_i}{\partial X_j}$ for $i \neq j$ are zero. Also, since the applied force is only in the z direction, symmetry requires $\frac{\partial x_2}{\partial X_2} = \frac{\partial x_1}{\partial X_1}$. Substituting these values into the energy equation reduces the number of possible solutions of Eq. 4.41 to a few real values. The largest of these real values corresponds to the only physically possible solution of Eq. 4.41. This result is accomplished without having to define nodes or tetrahedra. Using this result allows us to have Mathematica solve Eq. 4.41 for different input values of $\frac{\partial x_3}{\partial X_3}$. This is done in the following way:

1. A value of $\frac{\partial x_3}{\partial X_3}$ is chosen. Since we are dealing with a unit cube, $\frac{\partial x_3}{\partial X_3} = z$, the z component of the node along the z-axis. Set $\frac{\partial x_i}{\partial X_j} = 0$ for $i \neq j$ in ϵ. Substitute $\frac{\partial x_1}{\partial X_1}$ for $\frac{\partial x_2}{\partial X_2}$ values in ϵ.

2. The forces in the x and y directions are zero, so $\frac{\partial \epsilon}{\partial (\partial x_1 / \partial X_1)}$ is zero.

The chosen value of $\frac{\partial x_3}{\partial X_3}$ is substituted into $\frac{\partial \epsilon}{\partial (\partial x_1 / \partial X_1)} = 0$ and $\frac{\partial x_2}{\partial X_2}$ is solved for. Three solutions usually result, a negative, a positive, and a zero solution. Taking the positive value of the solution for $\frac{\partial x_2}{\partial X_2}$ finds the one physical solution. We now have all the values of $\frac{\partial x_i}{\partial X_j}$ and can solve for $\frac{\partial \epsilon}{\partial (\partial x_3 / \partial X_3)}$, which is the applied force per original area. Since the original area is 1, this is also the force applied to the surface.

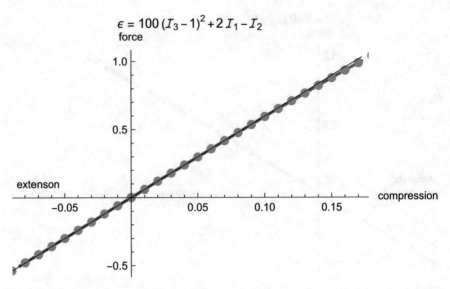

Fig. 8.5 Superimposed upon the results of Fig. 8.4 are dots showing the analytic solution of the cube under compression. This is an example of strain softening because at large forces (greater than 0.14) the compression is less than linear compression

From a set of input values of $\frac{\partial x_3}{\partial X_3}$ and calculated applied force values, a plot of compression versus force can be made and superimposed on the results of the node solution in Fig. 8.4; the results confirm the numerical solution results as shown in Fig. 8.5.

This testing is repeated for two additional energy functions as shown in Figs. 8.6 and 8.7.

A warning is necessary here. The energy function chosen should be validated against experiments over the entire range of its application. If this is not done, the results of the simulation may be faulty. For example, Fig. 8.8 shows the result of increasing the force applied to the linear energy function beyond that shown in Fig. 8.5. The result is that at about 30% compression, the simulation deviates from a linear trend, and after about 35%, it dramatically changes from the linear trend. Thus, the linear model energy function in Fig. 8.5 should not be used if any portion of the material is to be compressed more than 30% of its original size. This is not a limit of the basic equations but rather is a limit to the applicability of the chosen energy function in Fig. 8.5.

The first energy model shown in Fig. 8.2 is better behaved than the one in Fig. 8.8. Because it is more stable over a larger range of deformations, it may be a better model for those simulations requiring large extensions or compressions (see Fig. 8.9).

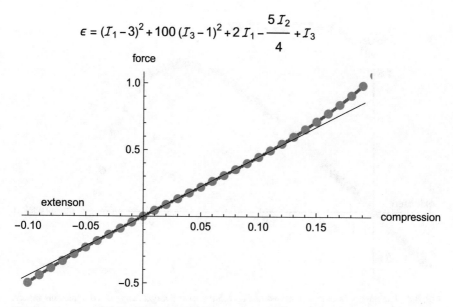

$$\epsilon = (I_1 - 3)^2 + 100\,(I_3 - 1)^2 + 2\,I_1 - \frac{5\,I_2}{4} + I_3$$

Fig. 8.6 An example of strain hardening. For large forces the compression is less than the linear compression

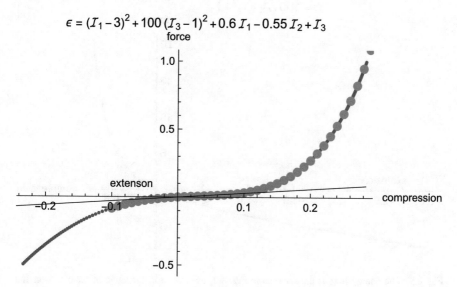

$$\epsilon = (I_1 - 3)^2 + 100\,(I_3 - 1)^2 + 0.6\,I_1 - 0.55\,I_2 + I_3$$

Fig. 8.7 This case of initially very small Young's modulus followed by a parabolic increase in strength may seem unusual, but it is characteristic of many biological materials

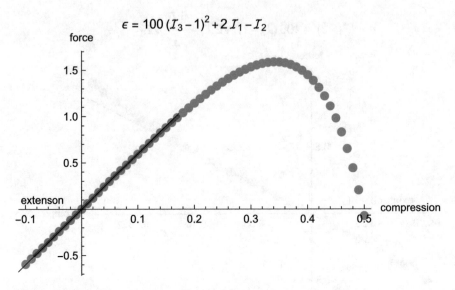

$$\epsilon = 100\,(\mathcal{I}_3 - 1)^2 + 2\,\mathcal{I}_1 - \mathcal{I}_2$$

Fig. 8.8 The energy here is that shown in Fig. 8.5 but over a larger range of compression values. Failure of this energy model occurs if the compression is more than about 30%

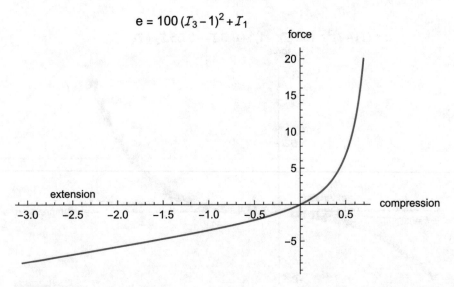

$$e = 100\,(\mathcal{I}_3 - 1)^2 + \mathcal{I}_1$$

Fig. 8.9 The energy here is the same as in Fig. 8.2, but over a larger range of compression and extension values

Cylindrical Compression Test

In this section, many nodes are used to simulate compressing a cylinder. This simulation will not be very physical, however. It is designed to test the computer code and make sure that the code produces the same results with many grids that were produced with a single grid in the last section. To make a comparison to the single grid example, the cylinder must be free to expand laterally without interference as a compressive force is applied. Even though this does not correspond to a physical experiment, it is a necessary exercise, so that the many-grid model may be tested and compared against the results of the analytic calculations of the last section.

The simulation to be used here consists of 408 nodes defining 1680 tetrahedra. Figure 8.10 shows the cylindrical geometry and the nodes used in the simulation.

The boundary conditions used to solve the compression of the cylinder are the same as those used in the last section to solve a single grid of four nodes. The same process is applied for the simulation: A force is applied to the top nodes of the cylinder. An energy function is defined. The simulation is carried out with a number of steps. The compression versus the force at each step is plotted. The final height of the cylinder is computed as the difference in the z values of the top and bottom nodes after deformation. This value is divided by 2 (since the original cylinder is of height 2) and subtracted from 1 to get the fractional displacement to compare to the single grid results. The applied force is increased linearly in five steps. The force is divided by the initial area of the top of the cylinder to compare this with the results of the single grid study. These results are shown for two different energy functions in Figs. 8.11 and 8.12.

These simulations show a good match to the analytic results of section "Single grid study". The final geometry of one of the deformations is shown in Fig. 8.13. Note how the cylinder uniformly expands as it is compressed. This is because all the nodes have been allowed to move freely in the radial direction.

Fig. 8.10. Outline of simulated cylinder showing the 408 nodes which make up the simulated material

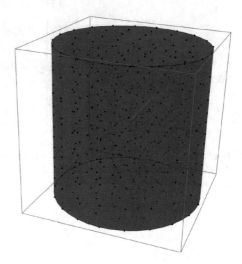

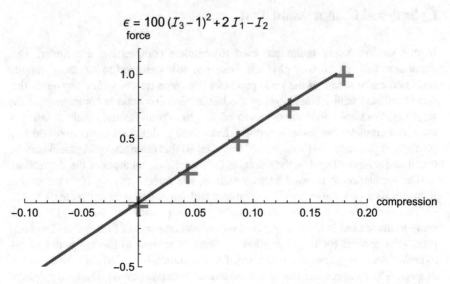

Fig. 8.11 The compression versus the force is plotted for the energy function shown. The red line is the result for a single grid calculation. The blue + signs show the results of the cylindrical compression with many grids

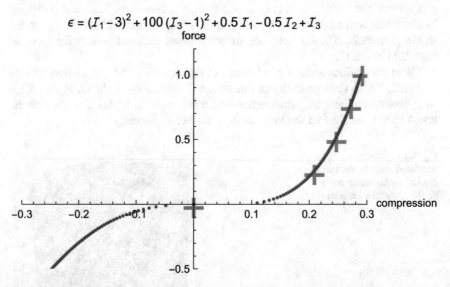

Fig. 8.12 The compression versus the force is plotted for the energy function shown. The red line is the result for a single grid calculation. The blue + signs show the results of the cylindrical compression with many grids

Fig. 8.13 The initial
cylinder in darker blue and
the final condition of the
cylinder in lighter blue, after
a maximum compression of
about 80% of the original
height of the cylinder

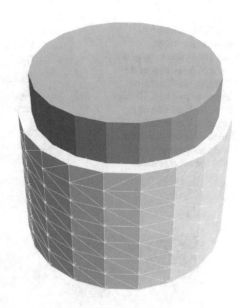

Real Compression

This simulation changes the compression from an ideal material where the top and bottom surfaces are free to move laterally to a more realistic case where the top and bottom points are fixed. This case is more realistic, because the force needed to compress the material increases the normal force on the top and bottom of the material. Since static friction is proportional to the normal force between two surfaces, this compressive force increases the friction, which prevents the top and bottom of the material from expanding laterally. Note that for this energy function, the multiple of $(I_3 - 1)^2$ is about 100 times the multiple of I_1 and I_2 so that the material is only slightly compressible as describe in section "Simulating incompressible materials" in Chap. 7. As a result, the material needs to expand laterally to conserve volume, so it "bulges" on the sides (see Fig. 8.14 for an example). The simulation producing Fig. 8.14 used the energy function used in Fig. 8.11.

Figure 8.15 shows that the "Dragon Skin" material used in the experiment in Chap. 10 responds similarly under compression.

A more standard compression experiment is one carried out on a cylinder instead of a cube of material. A simulation of this is shown in Fig. 8.16.

Just as experimental compression is not the same as the single grid simulation, so extension is not the same either. To extend a material, it must be attached to the object that is doing the pulling. Usually, a plate is glued to the top and bottom of a test material, and the top plate is then pulled to extend the material. The glue prevents the top and bottom of the material from moving laterally. The result of stretching is that the sides of the cylinder "squeeze" in (Fig. 8.17) to conserve volume.

Fig. 8.14 Simulated
compression of an
incompressible cube of the
material properties of the
"Dragon Skin" material
described in Chap. 10. Blue
is original position of the
material. The whiter,
bulging region shows the
final configuration of the
cube after compression

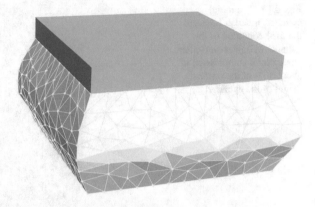

Fig. 8.15 Experimental compression of a cube of "Dragon Skin" material

Matching Fig. 2.1

This simulation will match the results of deforming Play-Doh as was shown in
Fig. 2.1. There is one "cheat" made in this simulation. The simulation is made of an
elastic material instead of a plastic material. An elastic material is one that stores
energy as forces do work on the material. This energy is returned to the outside world
as the forces are removed from the elastic material. When this happens, the material
returns to its original configuration after the forces are slowly removed. A plastic
material is permanently deformed by the forces on the material. It does not return to
its original shape, and it does not return the energy of deformation back to the
surroundings as the forces are removed but remains permanently deformed. In the
sixth example simulation of this chapter, we will learn a method of simulating plastic

Fig. 8.16 Simulation of
compression of a cylinder.
Blue is the original
configuration of the
material. The lighter blue
shows the material after
compression

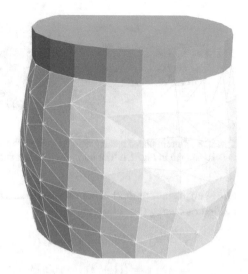

Fig. 8.17 Shown is the
original position of the
cylinder as a partially
transparent cylinder. The
final position of the material
after stretching it 15% in the
vertical direction is shown
"inside" the original
cylinder. Note how the sides
of the cylinder have moved
inward as a result of being
pulled

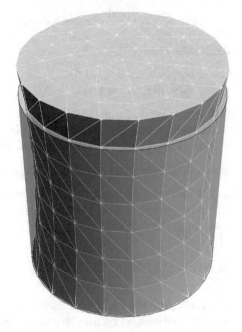

materials, but here an elastic simulation is used. This is not a great error if all that is
desired is the geometry of the final figure. If desired, the material can be re-gridded
as in the sixth simulation in this chapter to produce a plastic response.

The Play-Doh in Fig. 2.1 was pulled, and the ends rotated about the z-axis and
about the y-axis. To simulate this, one of the basic subroutines was modified to allow
for rotations as well as extensions and compression. With these two changes

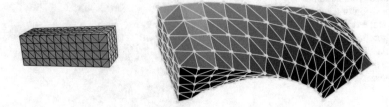

Fig. 8.18 The simulated rectangle on the left was "twisted" on the ends to match the deformation of the clay shown in Fig. 2.1. The figure on the right shows the result of the simulation

(allowing rotations and assuming the material to be elastic), the simulation was carried out using the following:

$$\mathbf{pg} = 0 \ (\text{i.e.no gravity})$$

$$\epsilon = (I1 - 3)^2 + 100(I3 - 1)^2 + 2I1 - 5/4I2 + I3$$

$$\mathbf{nsteps} = 10$$

Initial cube corners:

Lower left: $(-0.5, -0.5, -1.5)$
Upper right: $(0.5, 0.5 \text{ m } 1.5)$

Bottom nodes held fixed
    An initial rectangular distribution of nodes
    Top nodes stretched to 2.0 and rotated about z: $80°$ and rotated about y: $-55°$

area $= 0.02$ (this value sets the number of nodes; here 325 nodes resulted)

The results of the simulation are shown in Fig. 8.18 with the final figures rotated $90°$ about the x-axis. This should be compared visually to Fig. 2.1.

Bending Paper

Many thin materials can be simulated without a full 3D simulation. Paper, thin metal sheeting, and human skin may be more easily simulated with a 2D surface embedded in three dimensions, rather than a full 3D simulation. To simulate a surface embedded in three dimensions, two energies of deformation are required. The first is the energy of deformation in 2D. The second is the bending of the surface into the third dimension. These two energies of deformation must be defined before the simulation can proceed.

The 2D surface is defined by nodes located in the surface that have been triangulated so that the surface is made up of a set of nonoverlapping triangles. To define the energy associated with the expanding or shrinking or shearing of the 2D

Fig. 8.19. Normal unit vector $\widehat{v}_3$ perpendicular to the 2D surface defined by p_1, p_2, and p_3

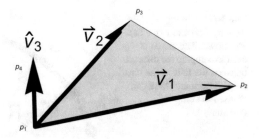

surface itself, define two vectors, $\vec{v}_1$ and $\vec{v}_2$, using the three corner nodes of each triangle $\vec{p}_1$, $\vec{p}_2$, and $\vec{p}_3$. Next, define a unit vector, $\widehat{v}_3$, which will always be perpendicular to the surface.

$$\begin{aligned}
\vec{v}_1 &= \vec{p}_2 - \vec{p}_1 \\
\vec{v}_2 &= \vec{p}_3 - \vec{p}_1 \\
\widehat{v}_3 &= \vec{v}_1 \times \vec{v}_2 / |\vec{v}_1 \times \vec{v}_2|
\end{aligned} \tag{8.1}$$

Equation 8.1 insures $\widehat{v}_3$ a unit normal to the surface as shown in Fig. 8.19.

The vector $\widehat{v}_3$ now defines a fourth point, $\vec{p}_4$ located at $\vec{p}_1 + \widehat{v}_3$, so that we now have four points $(\vec{p}_1, \vec{p}_2, \vec{p}_3, \vec{p}_4)$, which in turn define a tetrahedron for each triangle – just like in 3D. But now the volume of the tetrahedron is always proportional to the area of the surface since $\widehat{v}$ is a unit vector. The energy associated with deformations in the plane can now be calculated using

$$\text{energy} = f(\mathcal{I}_1, \mathcal{I}_2, \mathcal{I}_3) \tag{8.2}$$

In the problem to be studied in this section, the deformations in the plane of the surface are small compared to the bending of the surface, so the change in energy within the surface can be approximated using the linear energy formula

$$\text{energy} = a(\mathcal{I}_1 - 0.5\,\mathcal{I}_2) + b(\mathcal{I}_3 - 1)^2 \tag{8.3}$$

To calculate the bending energy, Kirchhoff's theory of bending is used. This theory of bending says the energy of bending of a rod is a function of the curvature squared, i.e.,

$$\text{energy} = \text{const.}\kappa^2. \tag{8.4}$$

The constant in Eq. 8.4 is proportional to Young's modulus times $r^4/4$, where r is the rod radius. Since r is small for a thin rod, this constant is always much less than the other constants of energy, i.e., "a" and "b" in Eq. 8.3. In our case, we will set the constant in Eq. 8.4 to a value much less than the other energy constants.

Fig. 8.20 The unit tangent
vectors, $\widehat{T}(s)$ and $\widehat{T}(s + \mathrm{ds})$,
are shown. The arc ds is
used to compute the distance
between the tails of the unit
tangent vectors. These can
be used to compute
curvature, κ

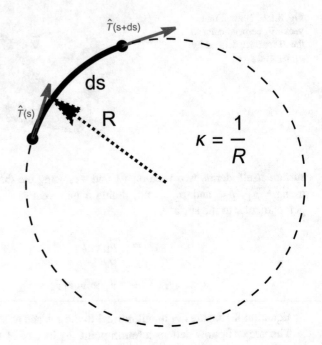

Curvature (which is 1/R for a circle of radius R) can be expressed in a number of
different ways, but the following is most useful for this application (Larson, Robert
P., Edwards, Bruce H. and Heyd, David E. Calculus with Analytic Geometry,
Seventh Ed, Houghton Mifflin Company, 2002, pg 823) :

$$\kappa = \frac{1}{R} = \left|\frac{d\widehat{T}}{ds}\right| = \lim_{ds\to 0}\left|\frac{\widehat{T}(s + ds) - \widehat{T}(s)}{ds}\right|, \tag{8.5}$$

where κ is the radius of curvature, $\vec{T}$ is a vector tangent to the circle, ds is the arc
length along the circle, and $\widehat{T}$ is the unit vector pointing along the $\vec{T}$ direction
(Fig. 8.20).

This definition of κ can also be written as

$$\kappa = \left|\frac{d\widehat{n}}{ds}\right| = \lim_{ds\to 0}\left|\frac{\widehat{n}(s + ds) - \widehat{n}(s)}{ds}\right|, \tag{8.6}$$

where $\widehat{n}$ is a unit vector perpendicular to the rod, Fig. 8.21.

When the curvature rod is discretized, κ can be approximated as

$$\kappa \approx \frac{|\widehat{n}_1 - \widehat{n}_2|}{s_1 + s_2}. \tag{8.7}$$

Fig. 8.21 The unit normal
vectors, $\hat{n}(s)$ and $\hat{n}(s + ds)$,
are shown. The arc ds is
used to compute the distance
between the tails of the unit
normal vectors. These can
be used to compute
curvature, κ

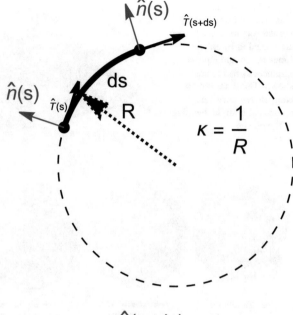

Fig. 8.22 The unit normal
vectors, $\hat{n}(s)$ and $\hat{n}(s + ds)$,
are shown for a discrete
representation of the circle.
The arc ds $= s_1 + s_2$ is used
to compute the distance
between the tails of the unit
normal vectors. Note that
now the unit normal vectors
are posted at the centers of
the line segments
approximating the circle.
These are used to
approximate the curvature,
κ, of the circle

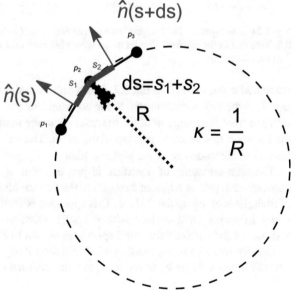

In Eq. 8.7, s_1 is the distance between the midpoint of the line segment defined by $\vec{p}_1$ and $\vec{p}_2$, and s_2 is the distance between the midpoint of the line segment defined by $\vec{p}_2$ and $\vec{p}_3$ as shown in Fig. 8.22.

Next, we need to apply Eq. 8.7 to a plane. For this simulation, triangular elements on a 2D surface are used. Between any two adjacent triangles, the curvature is

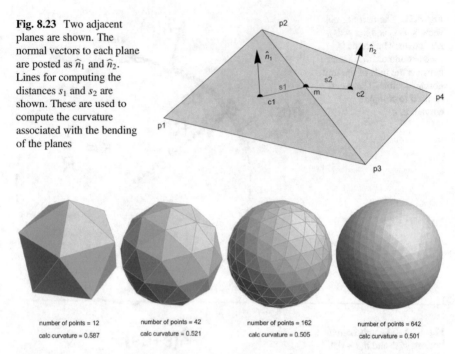

Fig. 8.23 Two adjacent planes are shown. The normal vectors to each plane are posted as $\hat{n}_1$ and $\hat{n}_2$. Lines for computing the distances s_1 and s_2 are shown. These are used to compute the curvature associated with the bending of the planes

number of points = 12   number of points = 42   number of points = 162   number of points = 642

calc curvature = 0.587   calc curvature = 0.521   calc curvature = 0.505   calc curvature = 0.501

Fig. 8.24 The "sphere" that is approximated in each case is of radius 2, with a curvature of exactly 0.5. Note that as the number of nodes in the sphere increases, the calculated curvature approaches the exact curvature

computed using Eq. 8.7. In Eq. 8.7, $\hat{n}_1$ is a unit vector normal to one of the triangles, and $\hat{n}_2$ is the unit vector normal to the adjacent triangle. The value of s_1 will be the distance from the midpoint of the triangle, $\vec{c}_1$, to the midpoint of the line segment of the common line between the two triangles, $\vec{m}$. The value of s_2 will be the distance from $\vec{m}$ to the midpoint of the adjacent triangle, $\vec{c}_2$ (see Fig. 8.23).

The total curvature of a surface is just the sum of all the κ values computed between each pair of adjacent triangles in the surface divided by the number of pairs of triangles making up the surface. This approach is verified in that as the number of nodes increases on a surface with a known curvature, the calculated curvature approaches the correct value for a sphere, as shown in Fig. 8.24.

Gravity will also be included in this simulation study. It is included in the energy per unit initial volume by introducing the gravitational energy as

$$\text{energy} = \sum_{i=1}^{N_p} \rho g Z_i, \tag{8.8}$$

where z_i is the z component of the location of the ith node in the simulation and N_p is the total number of nodes in the simulation. ρg is a user-set parameter.

The total energy per unit volume of a deformation is the combination of the energy from the deformation of the surface, the energy of bending of the surface, and the gravitational energy, i.e.,

$$\epsilon = a(\mathcal{I}_1 - 0.5\,\mathcal{I}_2) + b(\mathcal{I}_3 - 1)^2 + c\kappa^2 + \sum_{i=1}^{N_p} \rho g z_i \qquad (8.9)$$

All that remains is to set the constants a, b, c, and ρg.

For this application, the bending of a sheet of paper which is hanging off the edge of a desk is simulated. For this case, the forces involved will be too small to deform the paper in the plane of the paper. This will be simulated by making the constants "a" and "b" very large compared to "c". The force of gravity will also be small. The remaining variable is the ratio of "c" to the "ρg". The following parameters are chosen for the simulation:

$$\begin{aligned} a &= 10 \\ b &= 1000 \\ \rho g &= 0.1 \\ c &= 0.0035. \end{aligned} \qquad (8.10)$$

These parameters are fixed for all the simulations. The goal is to simulate a strip of paper 11" long and 2.2" wide hanging over a desk. Sixty-four nodes provided sufficient accuracy for the simulation of the paper strip. The maximum deflection of the paper from the horizontal was recorded. Figure 8.25 shows three examples of the resulting geometry.

Nine different amounts of overhang were measured, both experimentally and from simulations. Figure 8.26 shows a plot of the maximum vertical deflection as simulated and as experimentally measured. The deflection and the overhang are given in units of a fraction of the total length of the paper. These results indicate that using the model of curvature describe here is sufficient for the accuracy of this experiment. If this were not the case, a more complicated function of energy might be required.

Extrusion Study

Extrusion is used to create such varied products as car bumpers, weatherstripping, and macaroni. Probably the most widely known processes are Play-Doh extrusion (Fig. 8.27) and cake decorating (Fig. 8.28). A material is extruded when it is forced through a smaller opening than its cross section.

For this simulation example, the simulation is done in two dimensions. Elastic materials return to their original shape when applied forces are removed; plastic materials do not. After some initial deformation, they are permanently deformed by applied forces.

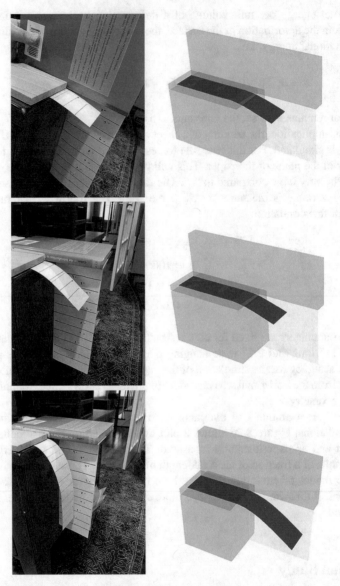

Fig. 8.25 The left column shows photos of the bending paper experiment. The right column shows the simulated results where the blue strip is the simulated strip of paper

To simulate a plastic response using Eq. 6.30, the material region is re-gridded after each step in the simulation process. That is, after minimizing Eq. 6.30 for a small displacement, a new grid is created, re-setting the energy to zero in the new grid before proceeding to minimize Eq. 6.30 on the new grid for the next step. This is repeated over and over again until the final deformation is produced.

Since ϵ is used in this simulation only for small deformations, the following linear ϵ is used

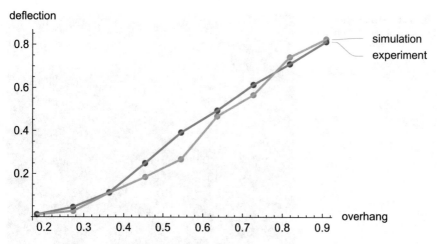

Fig. 8.26 This plot compares the maximum vertical deflection of the paper in the experiment and simulation plotted against the fraction of paper that was not over the desk, i.e., the overhang

Fig. 8.27 Play-Doh is extruded through a press

$$\epsilon = a(\mathcal{I}_1 - 1/2\,\mathcal{I}_2) + b(\mathcal{I}_3 - 1)^2 \qquad (8.11)$$

In this description of energy per unit volume, the value of "b" mainly controls the compressibility of the material, and the value of "a" mainly controls the shear. In this extrusion simulation, the piston location is set. Thus, no matter what the magnitude of shear force is required for deformation, it is applied in order to have the piston move. Thus, the shear parameter "a" is of little consequence in the shape of the extruded material. The material simulated is chosen to be essentially incompressible,

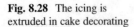

Fig. 8.28 The icing is extruded in cake decorating

so "b" is set to a large number compared to "a" (for this case, a = 1, b = 200). Gravity is considered negligible in the simulation (ρg = 0.0001). The only remaining parameter for determining material properties is the step size. Step size is set by the elastic limit of the material. Almost all plastic materials are elastic up to some deformation amount. This amount of deformation is called the elastic limit of the material (Fig. 8.29). In the simulation there are 20 steps, and the total deformation is 80%, so that each step is 4% of the total deformation and the elastic limit is modeled as 0.04.

The simulation begins with an orderly set of nodes. This is done so that the color of the material can be uniformly applied to the grid nodes. The orderly grid is quickly changed after a deformation of only a few steps, so that the starting grid has little effect on the final deformation. The top nodes are moved downward with each step. Each step re-grids. The extrusion proceeds until 80% of the material is extruded.

At step number 11 of 20, the distribution of the rate of energy loss is shown (Fig. 8.30). This is computed in the simulation by calculating the energy per unit original area in each triangular region just before re-griding is done. Since the energy of each triangle is set to zero after the re-griding, the calculated energy is the energy

Fig. 8.29 Force and extension plot of material which has yielded at the maximum deformation before re-griding

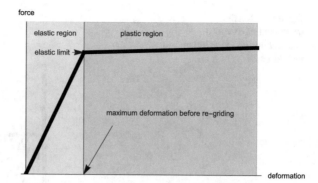

Fig. 8.30 The distribution of the rate of energy loss during extrusion. Red, most lost energy; green, middle lost energy; blue, least energy loss

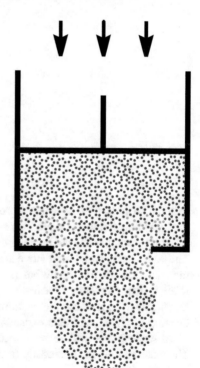

lost. The value is posted at the center of each triangle and displayed with a color map of blue being the least energy loss, green being in the middle, and red being the highest energy loss. The most lost energy corresponds to the most deformed regions of the extruded material. Not surprisingly this takes place at the edges of the opening. The amount of energy lost is mostly turned into heat energy. For soft extrusion like Play-Doh and cake frosting, there is very little change in temperature due to this heat loss, but in metal extrusion, the heat generated can change the temperature of the

Fig. 8.31 Relative
movement of extruded
material during extrusion at
step 11 of 20

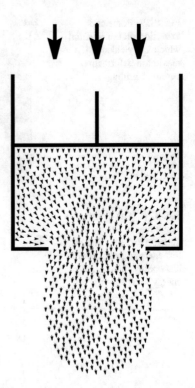

material so much that the material properties of the metal may change (e.g., melt) and
the high temperatures can produce stress on the die (the opening of the extruder).

At step number 11, the movement of the individual nodes is shown in Fig. 8.31.
The nodes in the simulation move in response to the motion of the piston pushing the
material through the die. What is shown is an arrow at each node showing the
relative displacement of each node just before the re-griding takes place in step
number 11. The material moves from the center of the un-extruded material into the
opening of the die and then expands into the extruded material.

At step number 20, the movement of the individual nodes is shown in Fig. 8.32.
The material now moves mostly from the sides of the un-extruded material into the
opening of the die.

The location of the nodes as they move are tracked. The starting and final position
of each node are saved at each time step. At the end of each time step, the material
region is re-gridded with a new set of nodes. A color is assigned to each node at the
initial time step. The color of the new nodes is assigned as the color of the nearest
displaced node in the previous time step. This process is repeated over and over for
all 20 time steps. At step 1, step 11, and step 20, the color distribution is shown in
Fig. 8.33. Notice that the gray material has been pushed into the core of the extrude
material. This same result can be seen qualitatively in Fig. 8.34, where Play-Doh has
been extruded with the initial volume being alternating orange and purple layers.

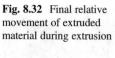

Fig. 8.32 Final relative movement of extruded material during extrusion

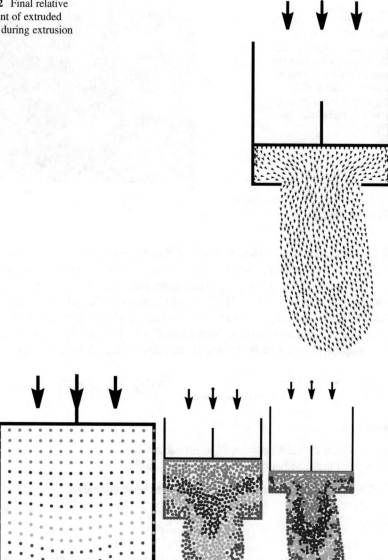

Fig. 8.33 Color bands as they are distorted during extrusion. Show is the simulation at steps 1, 11, and 20

Fig. 8.34 Extrusion of
layers of Play-Doh showing
similar behavior to the
simulated result in Fig. 8.33.
The Play-Doh has been cut
open after the extrusion to
show the motion of the
originally parallel layers of
orange and purple in the
Play-Doh

Water Drop

A 2D simulation can be used to simulate the shape of a water drop from a slowly
leaking faucet. Start with a set of nodes defining a circle. Since water has little to no
shear strength for slow (quasi-static) processes, the deformation energy function will
contain only one term, $(I_3 - 1)^2$. This term constrains the volume of the material.
There is a surface energy due to the surface tension of the liquid, so we must add an
energy term dependent upon the "surface area", i.e., in 2D the line area of the outside
nodes and in 3D the area of the outside surface. In 2D, this term is proportional to

$$\sum_{\text{all surface line segments}} \left(\sqrt{\text{dlf.dlf}} - \sqrt{\text{dl.dl}} \right)^2, \tag{8.12}$$

where

dl = original distance between adjacent surface nodes
dlf = final distance between adjacent surface nodes

Gravity is also needed. It will be defined as proportional to

$$\sum_{\text{all nodes}} \rho g z_i, \tag{8.13}$$

where ρg = user-defined quantity and z_i is the z component of the ith node.
This makes the energy per unit volume to be defined as

$$\epsilon = a \sum_{\text{all nodes}} (I_3 - 1)^2 \overline{V}_i$$
$$+ b \sum_{\text{all surface line segments}} \left(\sqrt{\text{dlf.dlf}} - \sqrt{\text{dl.dl}} \right)^2 + c \sum_{\text{all nodes}} \rho g z_i V_i, \tag{8.14}$$

where $\overline{V}_i$ is the initial area of each triangle and V_i is the initial area associated with
each node.

Fig. 8.35 The simulation of
a water drop in 2D. The
circle of black dots shows
the positions of the nodes
before deformation. The
blue area is the final area of
the material after the
simulation. The triangles are
the final shape of the
triangles used in the
simulation

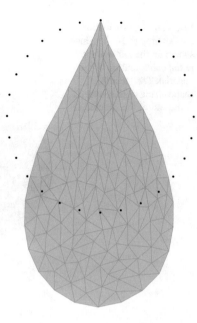

For this simulation, 129 nodes were used and

$$a = 10$$
$$b = 10$$
$$c = 1 \qquad\qquad (8.15)$$
$$\rho g = 0.01$$
$$\text{nsteps} = 5.$$

The results of the simulation in 2D are shown in Fig. 8.35.

The 2D calculation can be repeated in 3D. In 3D, ϵ is composed of the conservation of volume, gravity, and the surface area of the liquid:

$$\epsilon = a\sum_{i=1}^{N_g} (I_3 - 1)^2 \overline{V}_i + b\sum_{i=1}^{N_s} \overline{A}_i + c\sum_{i=1}^{N_n} \rho g z_i V_i, \qquad (8.16)$$

with

N_s = number or surface triangles
N_g = number of tetrahedra (grids)
N_n = number of nodes
$\overline{A}_i$ = current area of the ith surface triangle
$\overline{V}_i$ = volume of the ith tetrahedra (grid)
V_i = initial volume of the ith node.

Fig. 8.36 The simulation of
a water drop in 3D. The blue
sphere on the left was the
initial configuration of the
material. The "teardrop"
shape on the right is material
at the end of the simulation

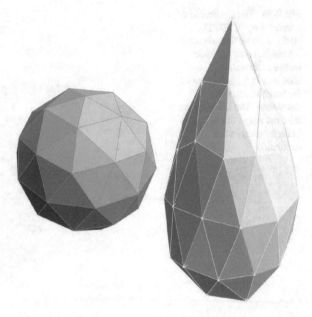

The first term constrains the volume of the drop, the second term is used to
minimize the current surface area, the third term defines the force of gravity. For this
simulation, 67 nodes were used and

$$
\begin{aligned}
a &= 1 \\
b &= 10 \\
c &= 1 \\
\rho g &= 1 \\
\text{nsteps} &= 3.
\end{aligned}
\tag{8.17}
$$

The results are shown in Figure 8.36.

Problems

Problem 1 Apply the following deformations: First $\mathcal{R}_1$ and then $\mathcal{R}_2$, to the
tetrahedron defined by the four points $(0, 0, 0)$, $(1, 0, 0)$, $(0, 1, 0)$, and $(0, 0, 1)$,
and find the final position of all four points, where

$$\mathcal{R}_1 = \begin{pmatrix} \cos\left(10^0\right) & \sin\left(10^0\right) & 0 \\ -\sin\left(10^0\right) & \cos\left(10^0\right) & 0 \\ 0 & 0 & 1 \end{pmatrix}$$

$$\mathcal{R}_2 = \begin{pmatrix} \cos\left(10^0\right) & 0 & -\sin\left(10^0\right) \\ 0 & 1 & 0 \\ \sin\left(10^0\right) & 0 & \cos\left(10^0\right) \end{pmatrix}.$$

Problem 2 Consider two triangles defined by the four points, (0, 0, 0), (1, 0, 0), (0, 1, 0), and (1, 1, .2). The first triangle contains the first three points, and the second triangle contains points 2, 3, and 4. Find the curvature defined by these two triangles.

Problem 3 Consider a right circular cylinder of height 10 cm with the axis along the z-axis and a radius of 0.2 cm. Compress this cylinder by 10% in the z direction with $\epsilon = 4\left(I_1 - 3\right)^2 + 100\left(I_3 - 1\right)^2$. Assume $\frac{\partial \epsilon}{\partial\left(\partial x_i / \partial X_j\right)} = 0$ for all i and j except $i = j = 3$. Use Eq. 4.41 to find the applied force. (Hint: This can be done using four points and one tetrahedron to represent the cylinder).

Problem 4 Suppose the elastic limit of a ductile material is 0.02, i.e., assume the material is elastic only up to a 2% deformation of the material. If we wish to quasi-statically deform the material by a factor of 40%, how many steps should the quasi-static process be divided into?

Problem 5 Show that normal vectors can be used instead of $\widehat{T}$ vectors for the curvature calculation:

That is, show $\left|\widehat{T}(s + ds) - \widehat{T}(s)\right| = |\widehat{n}(s + ds) - \widehat{n}(s)|$.

Problem 6 Find the formula for the surface area of a triangle defined by three points, p_1, p_2, and p_3.

Problem 7 Consider three nodes connected by two identical linear elastic springs with spring constants 1.0 N/cm and rest length of 1 cm. The nodes are constrained to move only in the $\pm z$ direction. These three nodes will be used to simulate two subregions of a uniform 1 cm $\times$ 1 cm $\times$ 2 cm rectangle of mass 3 g. This is arrangement is similar to Fig. 4.3. Node 1 (at $z = 0$) is fixed. Node 2 (initially at $z = -1$) has no external force on it except the two springs and gravity. Node 3 (initially at $z = -2$) has gravity acting down (-z direction) and a force of 0.2 N acting upward (+z) direction. (a) Write the J for this system, and (b) minimize J to find the final location of nodes 2 and 3.

Problem 8 Repeat problem 7, but this time use sum of forces equal zero to find the final position of the two nodes.

Chapter 9
The Invariants

Another Set of Invariants

Chapter 2 introduced the $\mathcal{F}$ matrix, Eq. 2.25, which defined how differential vectors are mapped from an undeformed state, dX_i, to a final state, dx_i. In Chap. 5, we learned that the energy of deformation of an isotropic body, ϵ, should be written in terms of the elements of $\mathcal{F}$ so that there are only three independent values – the I_i values. In this chapter, we will define another set of three invariants to define ϵ for isotropic bodies. In Chap. 10, we will explore how these alternative invariants help measure the material properties of elastic materials. In Chap. 12, we will also find them useful as we define anisotropic elastic materials. Even though these alternative invariants are useful, they will not be used in simulations because they require more computer time to calculate than the I_i invariants.

Any deformation defined by $\mathcal{F}$ can be written as a product of three matrices:

$$\mathcal{F} = \mathcal{R}_2 \, L \, \mathcal{R}_1, \tag{9.1}$$

where $\mathcal{R}_1$ and $\mathcal{R}_2$ are rotation matrices and L is a diagonal matrix,

$$L = \begin{pmatrix} L_{11} & 0 & 0 \\ 0 & L_{22} & 0 \\ 0 & 0 & L_{33} \end{pmatrix}. \tag{9.2}$$

The diagonal elements of L, L_{11}, L_{22}, and L_{33}, are another set of invariants of $\mathcal{F}$. These values, like the I_i values of Chap. 5, do not change regardless of the values of

Supplementary Information The online version contains supplementary material available at [https://doi.org/10.1007/978-3-031-09157-5_9].

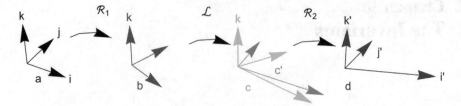

Fig. 9.1 Example of the mapping $\mathcal{F} = \mathcal{R}_2 \, L \, \mathcal{R}_1$. $\mathcal{R}_1$ rotates unit vectors at a into red unit vectors at b. L stretches rotated red unit vectors at b into green vectors at c. (The gray vectors in c' show that the original orientations of the initial unit vectors are only stretched by L, not rotated.) $\mathcal{R}_2$ rotates the green vectors at c into the final blue vectors at d. In each case, the vector sets have been displaced to be able to see the results of the mappings more clearly

$\mathcal{R}_1$ or $\mathcal{R}_2$ in Eq. 9.1. (The expression of $\mathcal{F}$ in Eq. 9.1 is similar to singular value decomposition (SVD), but SVD defines the $\mathcal{R}_i$ matrices as unitary matrices, which can include inversions – which are forbidden in material deformations.)

Equations 9.1 and 9.2 can be interpreted physically. $\mathcal{R}_1$ and $\mathcal{R}_2$ do not deform the material – they simply rotate it. $\mathcal{R}_1$ is the rotation *before* the deformation and $\mathcal{R}_2$ is the rotation *after* the deformation. The actual deformation of the material is determined by L. L stretches or compresses the material locally along the fixed coordinate directions.

Figure 9.1 shows the steps of the deformation L beginning with the unit vector, $\hat{i}$, $\hat{j}$, and $\hat{k}$. That is, start with $\vec{dX}$ as $\hat{i}$, $\hat{j}$, or $\hat{k}$. $\mathcal{R}_1$ operates first, rotating the corresponding unit vector. Next L stretches the space (and the rotated vectors) along the original $\hat{i}$, $\hat{j}$, and $\hat{k}$ axes. Finally, $\mathcal{R}_2$ rotates the resulting vectors to their final locations, $\hat{i'}$, $\hat{j'}$, and $\hat{k'}$, respectively.

In general, the mapping in Eq. 9.1 is nonunique. (For example, if $\mathcal{R}_2 L \mathcal{R}_1$ is a mapping, and so is $(-\mathcal{R}_2) L (-\mathcal{R}_1)$.) The mapping given in Eq. 9.1 can be made to be unique by defining the following steps:

1. The diagonal elements in L are found by calculating the eigenvalues of the matrix C defined as

$$C = (\mathcal{F})^T \, \mathcal{F}. \qquad (9.3)$$

Since C is a symmetric matrix, its eigenvalues are real. There will be three eigenvalues, λ_1, λ_2, and λ_3. The square root of these values will correspond to the three diagonal elements in L. Square roots in general can be $\pm$, but always choose the + root so that the determinant of L will be greater than zero. Thus

$$L = \begin{pmatrix} \sqrt{\lambda_1} & 0 & 0 \\ 0 & \sqrt{\lambda_2} & 0 \\ 0 & 0 & \sqrt{\lambda_3} \end{pmatrix}, \qquad (9.4)$$

giving

$$\mathcal{L}_{11} = \sqrt{\lambda_1}$$
$$\mathcal{L}_{22} = \sqrt{\lambda_2} \qquad (9.5)$$
$$\mathcal{L}_{33} = \sqrt{\lambda_3}$$

with $\mathcal{L}_{ij} = 0$ if $i \neq j$.

2. To find $\mathcal{R}_1$, first find the eigenvectors of C as $\vec{v}_1$, $\vec{v}_2$, and $\vec{v}_3$, corresponding to the eigenvalues λ_1, λ_2, and λ_3, respectively. These eigenvectors can be nonunique, but computer algorithms generate unique values. Normalize the computer-generated eigenvectors by dividing each by its length,

$$\hat{v}_i = \frac{\vec{v}_i}{\sqrt{\vec{v}_i \circ \vec{v}_i}} \text{ for } i = 1, 2, \text{ and } 3 \text{ (no sum)} \qquad (9.6)$$

The vectors, $\hat{v}_i$, will be orthogonal to one another and of unit length.

3. We can construct a 3×3 matrix with $\hat{v}_i$ as the row vectors of the matrix. (The transpose of this matrix defines a rotation or inversion since a matrix constructed with v_i as column vectors will map $\hat{i}$ into $\hat{v}_1$, $\hat{j}$ into $\hat{v}_2$, and $\hat{k}$ into $\hat{v}_3$.) To insure that the matrix we construct is a rotation matrix (and not an inversion), test if $(\hat{v}_1 \times \hat{v}_2) . \hat{v}_3$ is greater or less than zero. If this value is less than zero, interchange the vectors $\hat{v}_1$ and $\hat{v}_2$, so that $\hat{v}_1$ becomes the original $\hat{v}_2$ and $\hat{v}_2$ becomes the original $\hat{v}_1$. Also interchange $\mathcal{L}_{11}$ and $\mathcal{L}_{22}$. If $(\hat{v}_1 \times \hat{v}_2) . \hat{v}_3$ is greater than zero, leave the $\hat{v}_i$ unchanged.

4. Put the normalized vectors from step 3 as row vectors in a matrix $\mathcal{R}_1$ in the same order as the eigenvalues in C, i.e.,

$$\mathcal{R}_1 = \begin{pmatrix} \hat{v}_{1x} & \hat{v}_{1y} & \hat{v}_{1z} \\ \hat{v}_{2x} & \hat{v}_{2y} & \hat{v}_{2z} \\ \hat{v}_{3x} & \hat{v}_{3y} & \hat{v}_{3z} \end{pmatrix}. \qquad (9.7)$$

5. Construct $\mathcal{R}_2$ as

$$\mathcal{R}_2 = \mathcal{F}(\mathcal{R}_1)^T \mathcal{L}^{-1}, \qquad (9.8)$$

where

$$(\mathcal{L})^{-1} = \begin{pmatrix} 1/\mathcal{L}_{11} & 0 & 0 \\ 0 & 1/\mathcal{L}_{22} & 0 \\ 0 & 0 & 1/\mathcal{L}_{33} \end{pmatrix}. \qquad (9.9)$$

These five steps define a unique mapping from the matrix $\mathcal{F}$ to $\mathcal{R}_2 \, \mathcal{L} \, \mathcal{R}_1$.

The stretch (or compression) of the material is defined by the three diagonal elements of L, $\sqrt{\lambda_1}$, $\sqrt{\lambda_2}$, and $\sqrt{\lambda_3}$. These stretches take place after the rotation of the material by $\mathcal{R}_1$. If any of these values is equal to 1, no change in the material is made along the corresponding direction. If the value is less than 1, the material is compressed along that direction. If a value is greater than 1, the material is stretched along that direction. So, for example, if $\sqrt{\lambda_3}=1.2$, the material will be stretched by 20% in the z direction. Similarly, if $\sqrt{\lambda_3}=0.8$, the material will be compressed by 20% in the z direction.

The invariants, $\mathcal{L}_{ii}$, are related to the earlier invariants, I_i, as follows:

$$
\begin{aligned}
I_1 &= \mathcal{L}_{11}{}^2 + \mathcal{L}_{22}{}^2 + \mathcal{L}_{33}{}^2 \\
I_2 &= \mathcal{L}_{11}{}^2 \mathcal{L}_{22}{}^2 + \mathcal{L}_{11}{}^2 \mathcal{L}_{33}{}^2 + \mathcal{L}_{22}{}^2 \mathcal{L}_{33}{}^2 \\
I_3 &= \mathcal{L}_{11}\, \mathcal{L}_{22}\, \mathcal{L}_{33}
\end{aligned}
\tag{9.10}
$$

If the deformations do not include rotations, the I_i values have a particular physical interpretation. I_1 is the length of a diagonal of a unit cube with sides parallel to the coordinate planes. I_2 is the sum of the squares of the final areas of each side of the unit cube. I_3 is the final volume of a unit cube as long as the reference coordinate system is a "right-handed" coordinate system (i.e., one in which $\widehat{i} \times \widehat{j} = \widehat{k}$, not $\widehat{i} \times \widehat{j} = -\widehat{k}$).

We choose to make physical measurements of a material without rotating the material. If this is done, the three $\mathcal{L}_{ii}$ values will completely define the deformation (i.e., $\mathcal{R}_1$ and $\mathcal{R}_2$ will be identity matrices.) By measuring the stored energy to accomplish each deformation, a complete mapping of the energy of deformation of the material can be made. Since the energy per unit original volume, ϵ, is a function of I_i, it is also a function of $\mathcal{L}_{ii}$, that is,

$$
\epsilon = f(\mathcal{L}_{11},\, \mathcal{L}_{22},\, \mathcal{L}_{33}).
\tag{9.11}
$$

The $\mathcal{L}_{ii}$ values are all positive and vary around 1 depending upon the amount of deformation of the material. (A negative value of $\mathcal{L}_{ii}$ would correspond to an inversion of the material, which is nonphysical.) We can "map out" the energy of deformation, Eq. 9.11, by stretching and compressing an isotropic body along the coordinate axes.

In particular, we wish to find the energy for every value of $\mathcal{L}_{11}$, $\mathcal{L}_{22}$, and $\mathcal{L}_{33}$. But we only need the region of $\mathcal{L}_{ii}$ space which will be encountered in the deformations we wish to simulate. So, for example, if we do not expect any region of the simulated material to be compressed by more than 20%, we only need to consider the region of space for $\mathcal{L}_{ii} > 0.8$. If the maximum extension to be encountered is 50%, then we only need values for the energy for $\mathcal{L}_{ii} < 1.5$. If both conditions apply, then we need to find the values for ϵ for

Fig. 9.2 We need to find the ϵ values within the space defined by $\mathcal{L}_{11}$, $\mathcal{L}_{22}$, and $\mathcal{L}_{33}$ by varying the values of $\mathcal{L}_{ii}$ expected in the deformation to be simulated. Here the minimum value for each $\mathcal{L}_{ii}$ is 0.13 and the maximum value is 2.679

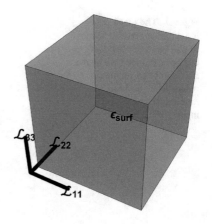

$$0.8 < \mathcal{L}_{11} < 1.5,$$
$$0.8 < \mathcal{L}_{22} < 1.5, \text{ and} \qquad (9.12)$$
$$0.8 < \mathcal{L}_{33} < 1.5,$$

A plot of this region of space is shown in Fig. 9.2.

For an incompressible material, the volume of any unit cube is fixed so that $\mathcal{L}_{11}$ $\mathcal{L}_{22}$ $\mathcal{L}_{33} = 1$. This forces

$$\mathcal{L}_{33} = \frac{1}{\mathcal{L}_{22}\mathcal{L}_{33}}, \qquad (9.13)$$

which defines a surface in the $\mathcal{L}_{ii}$ space. The combination of the two constraints in Eqs. 9.12 and 9.13 is shown in Fig. 9.3.

If the material is isotropic, symmetry further restricts the region of independent values of $\mathcal{L}_{ii}$ in that the energy is unchanged by any interchange of the $\mathcal{L}_{ii}$ values, i.e.,

$$\epsilon(\mathcal{L}_{11}, \mathcal{L}_{22}, \mathcal{L}_{33}) = \epsilon(\mathcal{L}_{11}, \mathcal{L}_{33}, \mathcal{L}_{22})$$
$$= \epsilon(\mathcal{L}_{22}, \mathcal{L}_{11}, \mathcal{L}_{33})$$
$$= \epsilon(\mathcal{L}_{22}, \mathcal{L}_{33}, \mathcal{L}_{11}) \qquad (9.14)$$
$$= \epsilon(\mathcal{L}_{33}, \mathcal{L}_{11}, \mathcal{L}_{22})$$
$$= \epsilon(\mathcal{L}_{33}, \mathcal{L}_{22}, \mathcal{L}_{11})$$

These conditions divide the surface into six regions. The value of the energy in any one region can be used to determine the energy value in any of the other five regions (Fig. 9.4).

For an incompressible, isotropic material, $\mathcal{L}_{33}$ is a function of $\mathcal{L}_{11}$ and $\mathcal{L}_{22}$, and ϵ is completely defined by $\mathcal{L}_{11}$ and $\mathcal{L}_{22}$. As a result, we can plot ϵ as a contour plot $\epsilon=$

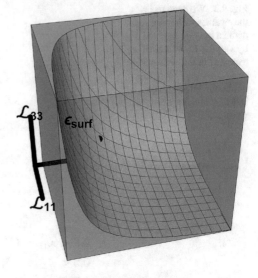

Fig. 9.3 The plane shows the range of possible values of the $\mathcal{L}_{ii}$ values that have been constrained both by the maximum and minimum values of $\mathcal{L}_{ii}$ and the requirement of incompressibility

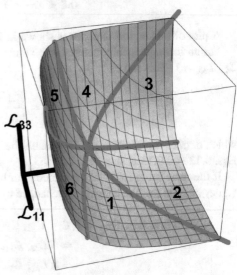

Fig. 9.4 Six regions of the ϵ surface any one of which will completely define ϵ due to symmetry

$f(\mathcal{L}_{11}, \mathcal{L}_{22})$. The range of $\mathcal{L}_{11}$ and $\mathcal{L}_{22}$ is defined by the range of expected deformations. Figure 9.5 shows the region defined by the constraints in Eqs. 9.12 and 9.13 as the gray semicircular region. Also shown are four regions within the plot which highlight compressive and extensional regions.

The shaded area shows the region bounded from Eqs. 9.12 and 9.13. The black point at line intersections of the vertical and horizontal lines corresponds to no deformation ($\mathcal{L}_{11} = 1$, $\mathcal{L}_{22} = 1$, and $\mathcal{L}_{33} = 1$). The four regions define compressional and extensional regions.

Fig. 9.5 Regions of ϵ plot
for incompressible material

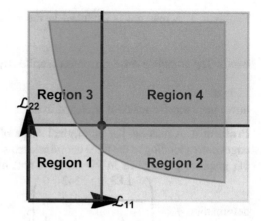

In region 1, $\mathcal{L}_{11} < 1$ and $\mathcal{L}_{22} < 1$, corresponding to compression along the x- and y-axes.

In region 2, $\mathcal{L}_{11} > 1$ and $\mathcal{L}_{22} < 1$, corresponding to extension along the x-axis and compression along the y-axis.

In region 3, $\mathcal{L}_{11} < 1$ and $\mathcal{L}_{22} > 1$, corresponding to compression along the x-axis and extension along the y-axis.

In region 4, $\mathcal{L}_{11} > 1$ and $\mathcal{L}_{22} > 1$, corresponding to extension along the x- and y-axes.

Equations 9.13 and 9.14 provide the basis for experimentally determining ϵ for a specific isotropic, incompressible material in Chap. 10. Once the $\mathcal{L}_{ii}$ values are obtained experimentally, Eq. 9.10 can be used to calculate the corresponding I_i values. We will find the $\mathcal{L}_{ii}$ values by performing experiments described in the next chapter.

Problems

Problem 1 Given $\mathcal{F} = \begin{pmatrix} .2 & .7 & 1 \\ 2 & 1.5 & .5 \\ 2.2 & 3.2 & 4.1 \end{pmatrix}$, find $\mathcal{R}_1$, $\mathcal{R}_2$, and $\mathcal{L}$ so that $\mathcal{F} = \mathcal{R}_2 \, \mathcal{L} \, \mathcal{R}_1$.

Problem 2 If a mapping is described by $\mathcal{F} = \begin{pmatrix} .2 & .7 & 1 \\ 2 & 1.5 & .5 \\ 2.2 & 3.2 & 4.1 \end{pmatrix}$, is the material compressed or extended in the x direction? By how much?

Problem 3 A material is stretched in the x direction and then rotated about the z-axis. The complete deformation is described by $\mathcal{F} = \begin{pmatrix} 1.88 & .342 & 0 \\ -0.684 & 0.940 & 0 \\ 0 & 0 & 1 \end{pmatrix}$. By how much was the material stretched in the x direction before it was rotated?

Problem 4 A material has an original shape of a cube 2 cm on a side, with three edges corresponding to the three coordinate axes. Draw a picture (in 3D if you have a 3D graphics package or in 2D (x, y) if not) of the material before and after the deformation $\mathcal{F} = \begin{pmatrix} 1.88 & .342 & 0 \\ -0.684 & 0.940 & 0 \\ 0 & 0 & 1 \end{pmatrix}$.

Problem 5 Equation 5.19 gives the I_1 values in terms of $\vec{a}$, $\vec{b}$, and $\vec{c}$, where $\vec{a}$, $\vec{b}$, and $\vec{c}$ are the column vectors of $\mathcal{F}$. Equation 9.10 gives the same I_i values in terms of $\mathcal{L}_{ii}$. Show that these are numerically the same for $\mathcal{F} = \begin{pmatrix} 1.1 & 1.1 & 2 \\ 6.2 & 1.4 & 3.2 \\ .4 & .6 & .3 \end{pmatrix}$.

Problem 6 Find $\mathcal{F}$ if a cube is compressed to 1/2 its size in the y and z directions (a) without rotations and (b) with a rotation of 20° about the z-axis before the compression.

Problem 7 Use a 3D graphics package to post the point $(1, 2, 0.5)$ on Fig. 9.4. Then plot the equivalent points defined by Eq. 9.14 on the same plot.

Problem 8 By how much does the volume of a material change when deformed by

$$\mathcal{F} = \begin{pmatrix} 1 & 1.2 & .3 \\ -.3 & .4 & .65 \\ 2.3 & 1.1 & 1.8 \end{pmatrix}?$$

Problem 9 Apply $\mathcal{F} = \begin{pmatrix} 1.2 & .2 & 0 \\ -.2 & .8 & .3 \\ .3 & .6 & 1 \end{pmatrix}$ to the three coordinate axes vectors. Use a 3D graphics package to plot before and after on the same plot. If you do not have a 3D graphics package, plot the x-y plane before and after on the same plot. Make the before coordinate set black and the after coordinate set red.

Problem 10 Show that for $\mathcal{F} = \begin{pmatrix} 1.1 & .1 & -.3 \\ .4 & .8 & .2 \\ .6 & .3 & 1.4 \end{pmatrix}$ that Eq. 9.10 is true. If you

have an algebraic solver, show that it is true for all $\mathcal{F}$. If you use an algebraic solver, the following algebraic solution to the cubic a $x^3 + $b $x^2 + $c x$+$d $= 0$ is useful:

$$x_k = -\frac{1}{3a}\left(b + \xi^k \text{ B} + \frac{\Delta_0}{\xi^k \text{B}}\right) \text{ for } k = \{0, 1, 2\}$$

with

$$\xi = \frac{-1 + \sqrt{-3}}{2}$$
$$\Delta_0 = b^2 - 3 \text{ a c}$$
$$\Delta_1 = 2 \text{ } b^3 - 9 \text{ a b c} + 27a^2 \text{ d}$$
$$\text{B} = \sqrt[3]{\frac{\Delta_1 \pm \sqrt{(\Delta_1)^2 - 4(\Delta_1)^3}}{2}}.$$

Problem 11 We know that $\mathcal{L}_{ii}$ spans the space of deformations for an isotropic body, because all deformations, $\mathcal{F}$, can be written as a function of the three $\mathcal{L}_{ii}$ terms (since $\mathcal{R}_1$ and $\mathcal{R}_2$ in Eq. 9.1 have no effect on deformations of isotropic bodies). Use an algebraic solver and Eq. 9.10 to solve $\mathcal{L}_{ii}$ as a function of I_i to show that I_1, I_2, and I_3 also span the space of all deformations for an isotropic body.

Chapter 10
Experiments

The objective of this chapter is to show how to take large deformation data from an elastic material and fit this data with an energy function, $\epsilon(I_1, I_2, I_3)$. The material used is called "Dragon Skin". A remarkably good fit is found to all experimental data from this material with only a single fit parameter. After the fit, a check is made by calculating the expected forces using Eq. 4.41 and comparing them to the forces measured in the experiment.

Experiments

Two different experimental measurements are described. The first applies compressional forces, and the second applies extensional forces. The material used in both cases is the elastic material Dragon Skin® 10 Very Fast, Platinum Silicone – Very Fast Cure, sold by Smooth-On, Inc., 5600 Lower Macungie Road, Macungie, PA 18062, phone 610-252-5800, www.smooth-on.com. The Dragon Skin "liquid rubber" comes in two bottles. The liquids in the two bottles are mixed in equal portions to form a solid, soft rubberlike material. The product claims to have a "pot life" of 4 min and a cure time of 30 min. I prepared the material forms in less than 4 min with no problems. I found the material still tacky after drying for 30 min, so I allowed the materials to cure overnight before using them in the experiments.

To form the compressive and extensional pieces, I prepare two molds using a 3D printer. One mold was cubical and had an inside measurement of 1.27 cm on each side; the other was flat and had inside measurements (length(L) x width(W) x thickness(T)) of 4.6 cm × 4.6 cm × 0.3 cm. I prepared the flat mold with "pegs" in the edges of the mold so that the material when formed would have holes recessed

Supplementary Information The online version contains supplementary material available at [https://doi.org/10.1007/978-3-031-09157-5_10].

Fig. 10.1 The blue square at the top left is the mold for the cubic Dragon Skin material. Bottom left is the cube after curing and being removed from the blue mold. The flat red square at the top right is the mold for the flat piece of Dragon Skin. Bottom right is the flat Dragon Skin material after removing it from the mold and adding hooks to attach to weights for stretching the material

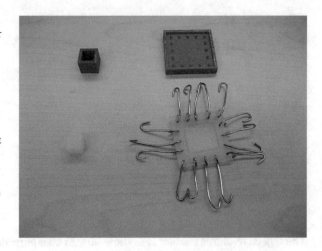

0.5 cm from each edge to accept hooks which I would use to stretch the thin piece of the Dragon Skin. The hooks were made from 12-gauge copper electrical wire (see Fig. 10.1).

I prepared the Dragon Skin material by pouring the liquids from the two bottles into a plastic container and stirring them with a toothpick to thoroughly mix them. I then poured the combined liquid material from the plastic container into each mold, stirring the material in the molds with a toothpick to remove air bubbles and to allow the material to completely fill each mold. I then used a flat piece of plastic to smooth off the top surface and remove any excess material from the top of the molds. I let the material cure overnight before removing them from the molds. I removed the Dragon Skin from the molds using a small screwdriver. I was able to slide the screwdriver along the sides of the material and separate it from the mold. Then I was able to pry the material from the mold without damaging it.

Compression Experiment

The cube was relatively easy to compress between my fingers (Fig. 8.15). I guessed from watching the cube deform under pressure that the volume of the cube was constant. To check this conclusion, I placed a piece of Lexan (clear plastic sheet 10 cm × 10 cm × 0.24 cm) on top of the cube. The Lexan sheet weighed only 25.5 g and did not compress the cube enough to measure. I measured width of the cube with just the Lexan sitting on it to be about 1.3 cm. Next, I applied pressure with my fingers on the sides of the Lexan to compress the cube vertically. After the compression, I measured the width of the cube (as seen through the Lexan) as 2.1 cm. The initial volume was $(1.27 \text{ cm})^3 = 2.05 \text{ cm}^3$; the final volume was $2.1 \text{ cm} \times 2.1 \text{ cm} \times 0.52 \text{ cm} = 2.29 \text{ cm}^3$. This is an increase in volume. I certainly would not expect the volume of the Dragon Skin material to increase under

Fig. 10.2 Bulge of Dragon Skin cube as it is compressed. Left is before compression. Right is after compression

Fig. 10.3 Display of stretching experiment

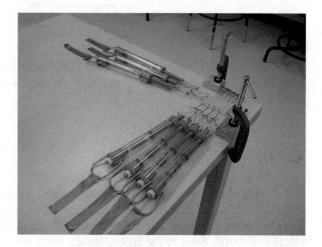

compression. I think the error in the width measurement occurred because after compression the cube was no longer perfectly a cube, but it bulged a bit on the sides (see Fig. 10.2). The width measurement made included the bulge and therefore was probably a bit too large. Because of this, I concluded that the material is not highly compressible. No material is completely incompressible, but with the results of this experiment, I will assume the volume of the Dragon Skin does not change appreciably with the forces to be applied during the following experiments.

Extension Experiments

The flat version of the material, shown on the right side of Fig. 10.1, was used to perform extensional experiments. The hooks on each of two sides of the material were connected to a nail in a block of wood, held by a clamp, so that these two bars were fixed. The hooks on the other two sides were attached to 250-g spring scales (Fig. 10.3). I drew a "square", which measured (length x width) 2.80 cm × 2.70 cm, on the flat Dragon Skin material with a pencil. This is the main region of the material to be deformed. I adjusted the spring scales so that the lines drawn on the flat material remained straight, forming a larger rectangle as the material was deformed. This insured more or less uniform deformation of the flat material. After making sure the lines drawn on the material remained straight, I taped down the spring scales so that they would not move while I recorded the force applied by each spring scale and the new size of the rectangle drawn on the surface of the material. The sum of the spring

Table 10.1 The force applied
and resulting length of the flat
Dragon Skin as a result of
pulling only in the length
direction

Step #	Length (cm)	Force (kg)
1	2.8	0.
2	3.1	0.19
3	3.3	0.31
4	3.92	0.46
5	4.11	0.56
6	4.25	0.635

scale forces was the total force, F, in each direction on the material. The measurement of the drawn rectangle is considered the region of the material that was uniformly deformed.

Extension in One Direction

Pulling only along the length direction, I measured the total applied force and the resulting length of the material. The results are found in Table 10.1.

To calculate the energy stored in the material due to the deformation, I need to calculate the work done by the applied force,

$$\text{work} = \int_{x_{\text{init}}}^{\vec{x}_{\text{final}}} \vec{F} \circ d\vec{x}. \tag{10.1}$$

The change in energy is the negative of the work done by the material or the positive work done by the applied force,

$$\Delta E = E - E_0 = \text{work} = \int_{\vec{x}_{\text{init}}}^{\vec{x}_{\text{final}}} \vec{F} \circ d\vec{x}. \tag{10.2}$$

I assume the initial energy, E_0, is zero. ϵ is the stored energy per unit initial volume. Divide E by the initial volume of the material,

$$\epsilon = \frac{E}{V_0}, \tag{10.3}$$

thus

$$\epsilon = \frac{1}{V_0} \int_{x_{\text{init}}}^{\vec{x}_{\text{final}}} \vec{F} \circ d\vec{x}. \tag{10.4}$$

In each case, the pull is in the direction that the material deforms. Thus

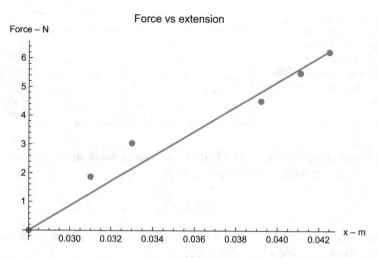

Fig. 10.4. Data from Table 10.1 and plot of the linear fit in Eq. 10.7

$$\int_{\vec{x}_{\text{init}}}^{\vec{x}_{\text{final}}} \vec{F} \circ \vec{dx} = \int_{\vec{x}_{\text{init}}}^{\vec{x}_{\text{final}}} \mid \vec{F} \parallel \vec{dx} \mid \cos 0^0 = \int_{\vec{x}_{\text{init}}}^{\vec{x}_{\text{final}}} F dx \qquad (10.5)$$

and finally

$$\epsilon = \frac{1}{V_0} \int_{x_{\text{init}}}^{\vec{x}_{\text{final}}} F dx. \qquad (10.6)$$

To evaluate this integral, we need continuous values of the applied force, F. A simple linear model is fit to the force (Fig. 10.4).

The fit equation is made by requiring the force, F, to be zero at the resting length of the material, x_0. A simple linear fit is made to the data. The result is as follows:

$$F = a(x - x_0)$$
$$a = 431.169 \text{ N/m} \qquad (10.7)$$
$$x_0 = 0.028 \text{ m}.x_0$$

Equation 10.7 is used along with the initial volume, L × W × T (2.8 cm × 2.7 cm × 0.3 cm), in Eq. 10.6 to calculate ϵ for each displacement. In this experiment, only the force along the length contributes to the stored energy. Even though the width of the material also changes (and thickness too), these do not contribute to the energy change because there is no external force in these directions. The results of using Eqs. 10.6 and 10.7 are shown in Table 10.2. Also shown in Table 10.2 are the $\mathcal{L}_{11}$ values calculated as

$$\mathcal{L}_{11} = \frac{\text{Length of material}}{\text{Original length of material}}. \qquad (10.8)$$

Step #	L_{11}	ϵ (10^3 N/m^2)
1	1.	0.
2	1.10714	0.855494
3	1.17857	2.37637
4	1.4	11.9237
5	1.46786	19.9853
6	1.51786	19.9853

Table 10.2 Energy stored in material per unit volume of original material (ϵ) as a function of dimensionless lengths (L_{11} = (length of the rectangle)/(initial length of the rectangle)

We need an equation of ϵ to be fitted to our data to use in simulations. A Taylor expansion of ϵ provides a starting place. To first order

$$\epsilon = a + b\, I_1 + c\, I_2 + d\, I_3, \tag{10.9}$$

where I_1 is a function of L_{ii}^2, I_2 is a function of $L_{ii}^2 L_{jj}^2$, and I_3 is a function of $L_{11} L_{22} L_{33}$. Therefore, to lowest order in L_{ii},

$$\epsilon = a + b\, I_1. \tag{10.10}$$

We have assumed in deriving Eq. 10.6 that $\epsilon = 0$ when no deformation has taken place, i.e., when $L_{11} = 1$, $L_{22} = 1$, and $L_{33} = 1$. In that case $I_1 = 3$, and Eq. 10.10 gives

$$0 = a + b(3) \tag{10.11}$$

or

$$a = -3\, b \tag{10.12}$$

and

$$\epsilon = b(I_1 - 3). \tag{10.13}$$

In terms of L_{ii}, I_1 is

$$I_1 = L_{11}^2 + L_{22}^2 + L_{33}^2. \tag{10.14}$$

From the compression experiment, I concluded that the material is incompressible. In that case,

$$L_{11} L_{22} L_{33} = 1 \tag{10.15}$$

Assuming symmetry,

$$L_{22} = L_{33} \tag{10.16}$$

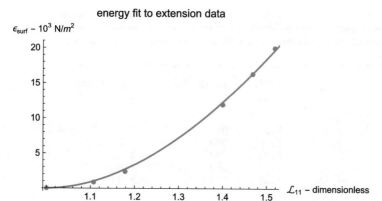

Fig. 10.5 Fit of Eq. 10.19 to data in Table 10.2

and

$$\mathcal{L}_{22} = \frac{1}{\sqrt{\mathcal{L}_{11}}}. \tag{10.17}$$

and

$$\mathcal{L}_{33} = \frac{1}{\sqrt{\mathcal{L}_{11}}}. \tag{10.18}$$

Equations 10.17 and 10.18 substituted into Eq. 10.13 gives

$$\epsilon = b\left(\mathcal{L}_{11}^2 + \frac{2}{\mathcal{L}_{11}} - 3\right). \tag{10.19}$$

which is our equation to fit the data in Table 10.2.

Figure 10.5 shows a fit of Eq. 10.19 to the ϵ data in Table 10.2. In that case, $b=3.16 \times 10^4$ N/m^2.

Extending in Two Directions

Next, I prepared extensions in both the length and width of the material. The results are found in Table 10.3.

The energy stored in an elastic material is only dependent upon the final state of the material. That is, the stored energy does not depend upon the path by which the final state was reached. Because of this, a two-step process can be used to calculate the stored energy. First the energy stored by a simple extension in the length is

Table 10.3 Size and force in both length and width directions. Note that the table is organized so that the length in each group is fixed while only the width is increased. Increasing both the length and width requires forces in both directions

step #	length cm	width cm	width force gm	length force gm
1	2.8	2.7	0	0
2	2.73	3.	200	120
3	2.72	3.32	390	235
4	2.77	3.63	530	235
5	2.73	3.88	670	350
1	3.02	2.59	0	270
2	3.02	2.85	195	275
3	3.02	3.14	360	275
4	3.02	3.43	540	395
5	3.02	3.84	715	475
1	3.43	2.55	0	400
2	3.43	2.74	205	410
3	3.43	2.98	335	485
4	3.43	3.185	480	490
5	3.42	3.61	626	515
1	3.72	2.51	0	485
2	3.72	2.72	190	500
3	3.72	3.02	380	510
4	3.72	3.08	530	625

Fig. 10.6 Steps in calculating energy when both width and length change. (**a**) Extend length, allowing width to change. (**b**) Extending width, holding length fixed

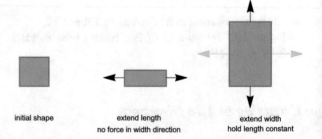

initial shape

extend length
no force in width direction

extend width
hold length constant

calculated. Next that energy is added to the energy stored by extending the width to the final state, holding the length fixed (Fig. 10.6).

In the first step, the integral of the force, Eq. 10.6, is used to calculate the energy stored in increasing the length to the value shown in column 1 of Table 10.3. In this step the width changes, but no width force is applied, so no energy is added due to the width deformation. Next, Eq. 10.6 is used to calculate the increase in energy as the width is increased from step 1 to step 5 in Table 10.3. But to compute this integral, a continuous width force is needed. Plotting the width forces versus the width using the data in columns 3 and 4 resulted in a nearly linear trend in each case. Thus a fit of this data to Eq. 10.7 provides the continuous width force needed. The "a" in Eq. 10.7 for each set of data is as follows:

$$\text{For the length of 2.8 cm} : a = 568.728 \text{ N/m}$$
$$\text{For the length of 3.02 cm} : a = 592.877 \text{ N/m}$$
$$\text{For the length of 3.42 cm} : a = 645.562 \text{ N/m}$$
$$\text{For the length of 3.72 cm} : a = 834.659 \text{ N/m}$$

The final energy is the sum of the stored energy from the first step plus the stored energy of the second step. The result for each calculation is found in Table 10.4. Also, in Table 10.4, L_{11} is calculated from Eq. 10.8; L_{22} is calculated as

Table 10.4 Energy per unit original volume that was stored in material (ϵ) as a function of dimensionless lengths (L_{11} = (length of the rectangle)/(initial length of the rectangle); L_{22} = (width of the rectangle)/(initial width of the rectangle); $L_{33} = \frac{1}{L_{11}L_{22}}$

step #	L_{11}	L_{22}	L_{33}	energy/V_0 (10^3 N/m^2)
1	1.03704	0.964286	1.	0.127076
2	1.01111	1.07143	0.923077	1.2555
3	1.00741	1.18571	0.837172	4.94672
4	1.02593	1.29643	0.751857	10.9713
5	1.01111	1.38571	0.713719	17.5851
1	1.11852	0.925	0.966529	1.23884
2	1.11852	1.01786	0.878355	2.1224
3	1.11852	1.12143	0.797233	5.19266
4	1.11852	1.225	0.729828	10.4614
5	1.11852	1.37143	0.651904	21.6615
1	1.27037	0.910714	0.864346	5.9528
2	1.27037	0.978571	0.804409	6.46658
3	1.27037	1.06429	0.739625	8.58429
4	1.27037	1.1375	0.692019	11.6915
5	1.26667	1.28929	0.612334	21.9438
1	1.37778	0.896429	0.809665	11.0678
2	1.37778	0.971429	0.747154	11.8793
3	1.37778	1.07857	0.672933	15.8538
4	1.37778	1.1	0.659824	17.0462

$$\mathcal{L}_{22} = \frac{\text{Width of material}}{\text{Original width of material}} \qquad (10.20)$$

and $\mathcal{L}_{33}$ as

$$\mathcal{L}_{33} = \frac{1}{\mathcal{L}_{11}\mathcal{L}_{22}}. \qquad (10.21)$$

Now we need an equation of ϵ to fit this data. Equation 10.13 is used again, but now we have Eqs. 10.21 and 10.22 instead of Eqs. 10.18 and 10.19 for $\mathcal{L}_{22}$ and $\mathcal{L}_{33}$. This gives

$$\epsilon = b\left(\mathcal{L}_{11}^{\,2} + \mathcal{L}_{22}^{\,2} + \frac{1}{\mathcal{L}_{11}^{\,2}\mathcal{L}_{22}^{\,2}} - 3 \right). \qquad (10.22)$$

Fitting Eq. 10.22 the data in Table 10.4 gives

$$b = 3.367 \times 10^4 \text{N/m}^2. \qquad (10.23)$$

Is this a "good" fit? One way to compare the fit to the data is to plot the energy from both the equation and the measured data on the same plot. Since $\mathcal{L}_{33}$ is a function of $\mathcal{L}_{11}$ and $\mathcal{L}_{22}$, ϵ is a function of only $\mathcal{L}_{11}$ and $\mathcal{L}_{22}$. Therefore, a contour plot can be used to compare the data to the fit. Before we do this, however, first post the location of the data points. Figure 10.7 plots the location of the energy values in Table 10.4 with $\mathcal{L}_{11}$ as the abscissa and $\mathcal{L}_{22}$ as the ordinate.

In addition to the plot of the raw data in Table 10.4, note that ϵ in Eq. 10.22 is symmetrical with respect to the $\mathcal{L}_{ii}$ values. Thus, there is no difference between ϵ ($\mathcal{L}_{11}$, $\mathcal{L}_{22}$, $\mathcal{L}_{33}$) and ϵ ($\mathcal{L}_{22}$, $\mathcal{L}_{11}$, $\mathcal{L}_{33}$). In other words, each data point posted at ($\mathcal{L}_{11}$, $\mathcal{L}_{22}$, $\mathcal{L}_{33}$) can also be posted at ($\mathcal{L}_{33}$, $\mathcal{L}_{11}$, $\mathcal{L}_{22}$). In fact, any permutation of the $\mathcal{L}_{ii}$

Fig. 10.7 The location of the points of measured data. The different regions correspond to compression in the x and y directions, extension in the x and compression in the y, extensions in the y and compression in the x, and extensions in the x and y directions

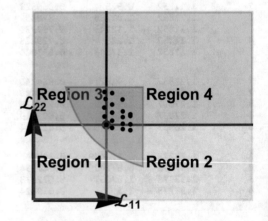

Fig. 10.8 Data points
which were measured in
black with data points
inferred in red

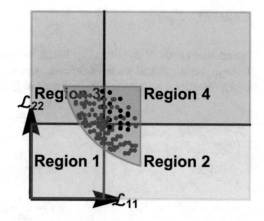

Fig. 10.9 A contour plot of
the measured data posted at
the locations in Fig. 10.8.
Also displayed in white are
the contour lines of the fit of
the original data to
Eq. 10.22

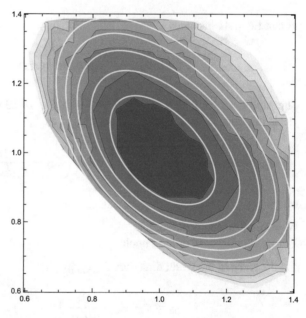

values must yield the same energy. Figure 10.8 posts all of these permutations on the
$\mathcal{L}_{11} - \mathcal{L}_{22}$ plot. The data from these few experiments covers the range of possible $\mathcal{L}_{ii}$
values quite well.

Each of the points in Fig. 10.8 has a corresponding measured energy value. These
values are used to create a contour plot of the energy data. The color regions in
Fig. 10.9 correspond to energy values. Super imposed on the color contour plot of
the data are constant contour values of the fit in white. Clearly, we have a good fit to
the energy.

Checking Forces

As another check of our fit to Eq. 10.22, we use Eq. 4.41 to compute the expected force in the length and width directions. Here the stretches are only in the length and width directions, so the $\partial x_i/\partial X_j = 0$ for $i \neq j$. Also

$$L_{11} = \frac{\partial x_1}{\partial X_1} \tag{10.24}$$

$$L_{22} = \frac{\partial x_2}{\partial X_2} \tag{10.25}$$

$$L_{33} = \frac{\partial x_3}{\partial X_3}, \tag{10.26}$$

so that Eq. 4.41 becomes

$$\mathcal{P}_{ii} = \frac{\partial \epsilon}{\partial L_{ii}} \text{ (no sum over } i\text{)}. \tag{10.27}$$

Choosing $i = 1$ for length and $i = 2$ for width, Eq. 10.22 substituted into Eq. 10.27 yields the force in the length direction, F_{length}, as

$$F_{\text{length}} = \mathcal{P}_{11} \; A_1 = \frac{\partial \epsilon}{\partial L_{11}} (W \times T) = 2 \; b \left(L_{11} - \frac{1}{L_{11}{}^3 L_{22}{}^2} \right) (W \times T), \tag{10.28}$$

where

W = initial width of the sample
T = initial thickness of the sample

and the force in the width direction, F_{width}, as

$$F_{\text{width}} = \mathcal{P}_{22} \; A_2 = \frac{\partial \epsilon}{\partial L_{11}} (L \times T)$$

$$= 2 \; b \left(L_{22} - \frac{1}{L_{11}{}^2 L_{22}{}^3} \right) (L \times T), \tag{10.29}$$

where L = initial length of the sample.

The numerical results are listed in Table 10.5.

As a further comparison, Figs. 10.10 and 10.11 plot Eqs. 10.28 and 10.29 along with the original data in Table 10.5.

These fits are not perfect, but they are quite good considering the fit of all this data was made using a single fit parameter, Equation 10.22.

Table 10.5 The length and width along with the calculated applied force in the width and length directions on the Dragon Skin material

step #	length cm	width cm	width force gram	length force gram
1	2.8	2.7	0.	0.
2	2.73	3.	220.564	105.756
3	2.72	3.32	399.301	188.479
4	2.77	3.63	538.518	248.671
5	2.73	3.88	634.985	287.075
1	3.02	2.59	- 8.42488	118.249
2	3.02	2.85	187.399	202.198
3	3.02	3.14	355.828	272.345
4	3.02	3.43	491.267	325.464
5	3.02	3.84	648.46	381.028
1	3.43	2.55	88.5488	342.386
2	3.43	2.74	217.722	387.833
3	3.43	2.98	350.987	433.284
4	3.43	3.185	446.578	464.251
5	3.42	3.61	609.897	508.992
1	3.72	2.51	129.558	464.848
2	3.72	2.72	261.641	505.62
3	3.72	3.02	411.944	549.778
4	3.72	3.08	438.164	557.097

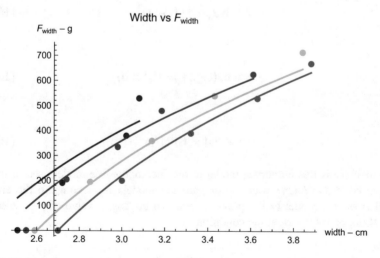

Fig. 10.10 Plot of computed force along the width as a function of width for different lengths (red, 2.8 cm; green, 3.02 cm; blue, 3.43 cm; purple, 3.72 cm)

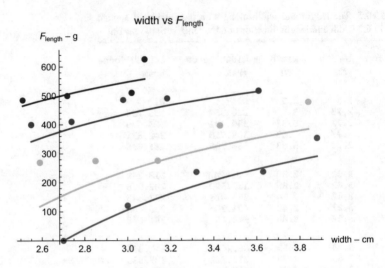

Fig. 10.11 Plot of computed force along the length as a function of width for different lengths (red, 2.8 cm; green, 3.02 cm; blue, 3.43 cm; purple, 3.72 cm)

Simulations

Following the discussion of section "Plots" in Chap. 7, we simulated Dragon Skin in Chaps. 7 and 8, using the following model of ϵ:

$$\epsilon = b\, I_1 + \gamma(I_3 - 1)^2, \tag{10.30}$$

with

$$\gamma \gg b, \ (e.g., \gamma = 100 \times b) \tag{10.31}$$

and

$$b = 3.4 \times 10^4 \mathrm{N/m^2}. \tag{10.32}$$

How close is this numerical model to our incompressible model used to fit our Dragon Skin data? One way to compare the models is to compare the energy function of this model to the previous one. To do this, a value of γ is selected: 100, 1000, or 10000. Next the condition

$$\delta \int_{V_0} \epsilon dV_0 = 0 \tag{10.33}$$

is imposed by setting

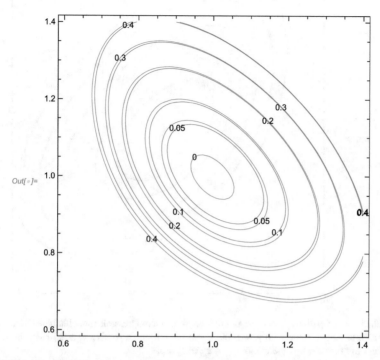

Fig. 10.12 A comparison of the energy function of the incompressible model in blue to the slightly compressible model in red with $\gamma = 100$. (Contour values are relative values.) For this value of γ, there is enough difference in the two models to see clearly the difference in their energy function plots

$$\frac{\partial \epsilon}{\partial \mathcal{L}_{33}} = 0 \qquad (10.34)$$

and solving for $\mathcal{L}_{33}$.

This value of $\mathcal{L}_{33}$ is substituted back into ϵ so that ϵ becomes a function of only $\mathcal{L}_{11}$ and $\mathcal{L}_{22}$. A contour plot is then made of $(\mathcal{L}_{11}, \mathcal{L}_{22})$. The results are found in Figs. 10.12, 10.13, and 10.14.

Note first that the contour lines are more similar as γ increases. Second note that the contours in extension (upper right quadrant of the plot) are slightly better than those in compression (lower left quadrant of the plot) for $\gamma = 100$. This difference all but disappears as γ is increased to 1000 and 10000. Also, less noticeable is that the center of the ellipse contour of the slightly compressible model is displaced slightly from $(\mathcal{L}_{11}, \mathcal{L}_{22}) = (1, 1)$. In other words there is a slight pre-stress on the material. This is shown in Fig. 8.2 in that the zero of the force-displacement curve is offset slightly from $(0, 0)$. The pre-stress becomes smaller as γ is increased.

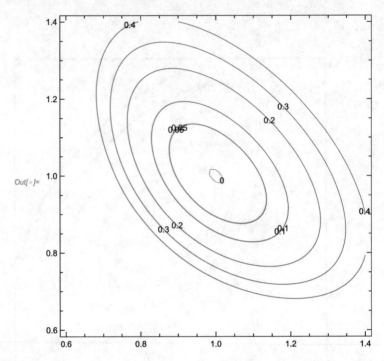

Fig. 10.13 A comparison of the energy function of the incompressible model in blue to the slightly compressible model in red with $\gamma = 1000$. (Contour values are relative values.) Here the red and blue lines are so close that the combination appear purple

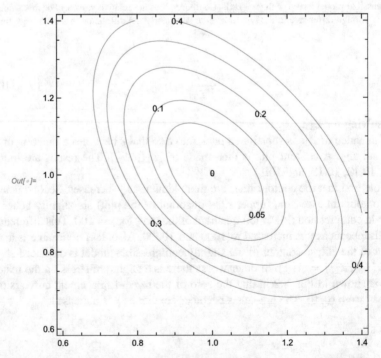

Fig. 10.14 A comparison of the energy function of the incompressible model in blue to the slightly compressible model in red with $\gamma = 10000$. (Contour values are relative values.) The blue lines are completely covered by the red contour lines because the models are so similar – no difference can be seen in the contours

Now the obvious question: Which model is the "correct" model: the incompressible model or the slightly compressible model? The answer is that with the data taken, we cannot tell. Both models are within the error of the measurements taken, and either model would be acceptable to make predictions within the accuracy of the measurements used to fit it. To distinguish between these models would require more accurate measurements.

This is a place to note that just as models should not be extrapolated beyond the range of the data, so also models should not be used to extrapolate to below the inherent measurement error used to define them. For example, if a model has been calibrated to 0.1 cm, it is very risky to expect the model to give correct results to 0.001 cm.

Problems

Problem 1 Calculate the width of an incompressible, symmetric cube 1.27 cm on a side if the cube is squeezed from the top and bottom and deformed from 1.27 cm to 0.52 cm. No forces are applied to the sides of the cube, and there is no friction on the top and bottom of the cube.

Problem 2 What is the width of the flat Dragon Skin material after it has been stretched to 3.8 cm in length?

Problem 3 Fit Eq. 10.7 to the force data given in Table 10.1 and find a. (Hint: units!)

Problem 4 Fit the data from Table 10.2 with a quadratic representation of L_{11} (i.e., a $L_{11}^2 + b\, L_{11} + c$). Then plot this relationship and Eq. 10.19 from $L_{11} = 0.01$ to $L_{11} = 2.0$ on the same plot. Which relationship is more likely to describe a real incompressible, symmetric material? Why?

Problem 5 Use Eqs. 10.6 and 10.7 to find the energy per unit initial volume when the material (original width $= 2.7$ cm, length $= 2.8$ cm, and thickness $= 0.3$ cm) has been stretched to 4.11 cm in the length direction. Check the answer using Eq. 10.19. (Remember the data and fit are only to $\pm 5\%$.)

Problem 6 If an energy datum for an incompressible material is posted at $L_{11} = 1.1$ and $L_{22} = 0.8$, list five other locations this same energy could be posted at.

Problem 7 Calculate the energy when the Dragon Skin has been stretched to a width of 3.0 cm and a length of 3.0 cm.

Problem 8 Show that the force at $L_{11} = 1$ is zero in Eq. 10.19.

Problem 9 Use Eqs. 10.6 and 10.7 to find the stored energy of the flat Dragon Skin material after a stretch of 1 cm in the length direction.

Problem 10 Use Eq. 10.27 and the energy in Eq. 10.22 to calculate the expected length force on the flat Dragon Skin material when the length is 3.02 cm and the width is 3.14 cm. Compare your results to the results in Table 10.5.

Chapter 11
Time-Dependent Simulations

So far, we have always sought the quasi-static solution to Eq. 4.45, where the acceleration of each node is zero. Now we wish to include acceleration in the numerical calculations. Equation 4.45 states that the sum of the forces acting on each small region of the material results in the acceleration of that small region. Since we are dealing with hyper-elastic materials, all forces on the material are conservative forces and can be written as the gradient of the energy of deformation. The total energy associated with a deformation was found in Chap. 6 to be described by Eq. 6.17. Using this energy, the sum of all the forces on each node, $\Sigma \vec{F}$, can be written as

$$\sum \vec{F} = -\vec{\nabla} E_{\text{tot}} \qquad (11.1)$$

where the gradient is with respect to the current position of the node, not the original position (i.e., small x, not capital X). That is

$$\vec{\nabla} = \left(\frac{\partial}{\partial x}, \frac{\partial}{\partial y}, \frac{\partial}{\partial z} \right) = \left(\frac{\partial}{\partial x_1}, \frac{\partial}{\partial x_2}, \frac{\partial}{\partial x_3} \right) \qquad (11.2)$$

The equation of motion for each node can be written in terms of components as

$$m \, a_i = - \frac{\partial}{\partial x_i} E_{\text{tot}}, \qquad (11.3)$$

where

m is the mass of the node
a_i is the ith component of the acceleration of the node

Supplementary Information The online version contains supplementary material available at [https://doi.org/10.1007/978-3-031-09157-5_11].

x_i is the ith component of the current position of the node
E_{tot} is the total energy of the system given in Eq. 6.29.

Equation 11.1 says roughly that if we slightly displace the final position of a node without moving any other nodes, and calculate the change in energy, the ratio of the energy change to the node displacement is the negative of the total force on the node. The next section shows that Eq. 11.3 divided by the initial volume associated with a node is the same as Eq. 4.45. The equivalence of Eqs. 11.3 and 4.45 is important because Eq. 11.3 will be used in time-dependent numerical simulations.

Example in 1D

The demonstration that Eqs. 11.3 and 4.45 are equivalent is a bit messy in 3D. To set the stage for 3D, this section will illustrate Eq. 11.3 for a small case of five nodes in one dimension without gravity. Let the nodes be connected by elastic springs with spring constants of 10. Assume the five nodes are initially located at points (0, 1, 2, 3, 4). A force of 2.1 is applied to node 5 to compress the nodes closer together (i.e., $\vec{F}$ is in the $-$ x direction). Node 1 is held fixed (Fig. 11.1).

The energy stored in each spring is

$$E_i = 1/2 \ k((x_{i+1} - x_i) - (X_{i+1} - X_i))^2 \ = 5 \ ((x_{i+1} - x_i) - 1)^2 \qquad (11.4)$$

with x_i the position of the ith node.

Equation 6.10 gives

$$E_{tot} = 5((x_1 - x_1) - 1)^2 + 5((x_2 - x_1) - 1)^2$$
$$+5((x_3 - x_2) - 1)^2 + 5((x_4 - x_3) - 1)^2 \qquad (11.5)$$
$$+5((x_5 - x_4) - 1)^2 - (-2.1)(x_5 - X_5).$$

1   2   3   4   5   Before

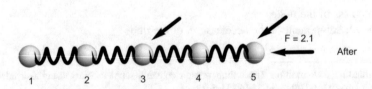

F = 2.1

After

1   2   3   4   5

Fig. 11.1 Five-node system before and after an applied force of 2.1. Node 1 is fixed. Node 5 has a force of 2.1 applied to it. Nodes 3 and 5 are the nodes used in the text to calculate forces after the deformation

The total force on node 3, F_3, can be calculated using Equation 11.1 as

$$F_3 = -\nabla_{x_3} E_{\text{tot}} = -\frac{\partial}{\partial x_3} E_{\text{tot}}. \qquad (11.6)$$

Note that the derivative is with respect to x_3 and all the energy terms are zero except for the third and fourth terms in Eq. 11.5. Thus

$$F_3 = -\nabla_{x_3} E_{\text{tot}} = -\frac{\partial}{\partial x_3} E_{\text{tot}} = -10((x_3 - x_2) - 1) + 10((x_4 - x_3) - 1), \quad (11.7)$$

and this gives an equation for the total force on node 3.

The total force on node 5 is

$$F_5 = -\nabla_{x_5} E_{\text{tot}} = -\frac{\partial}{\partial x_5} E_{\text{tot}} = -10((x_5 - x_4) - 1) + 2.1. \qquad (11.8)$$

Notation Alert

In the next section, notation is needed for both the component and the node number. The small letters i, j, and k will correspond to component designations, where i, j, and k vary from 1 to 3. Capital letters I, J, K, etc., will correspond to node number designations and will vary over the number of nodes.

Equivalence of Equations of Motion

This section will show that Eqs. 11.3 and 4.45 are equivalent if both sides of Eq. 11.3 are divided by the initial volume associated with each node. Begin with the gradient evaluated at the Pth node,

$$\vec{\nabla}_P E_{\text{tot}} = \vec{\nabla}_P \left(\sum_{I=1}^{N_v} \epsilon_I V_I \right) + \vec{\nabla}_P \left(\sum_{K=1}^{N_n} \rho_0 g(z_K - Z_K) \overline{V}_K \right)$$
$$- \vec{\nabla}_P \left(\sum_{J=1}^{N_b} \vec{f}_{\text{app}} \circ \left(\vec{x}_J - \vec{X}_J \right) A_J \right). \qquad (11.9)$$

Consider the last term first. For this term, only those nodes which have applied forces will give a nonzero value. For each node with an applied force,

$$\vec{\nabla}\left(\vec{f}_{\text{app}} \circ \left(\vec{x}_j - \vec{X}_j\right)A_j\right) = \left(\frac{\partial}{\partial x}, \frac{\partial}{\partial y}, \frac{\partial}{\partial z}\right)$$

$$\times \left(f_x(x - X)A_x + f_y(y - X)A_y + f_z(z - Z)A_z\right)$$

$$= \left(f_x A_x, f_y A_y, f_z A_z\right) = \left(F_x, F_y, F_z\right) = \vec{F},$$

$$(11.10)$$

where $\vec{F}$ is the applied force on a particular node.

The second term in Eq. 11.9 gives a value for each node, but only one term in the sum (the one corresponding to the Pth node) is nonzero, since the partial derivative with respect to every other node in the sum is zero. In addition, only the z direction is nonzero. That is, for the Pth node

$$\vec{\nabla}_P\left(\sum_{K=1}^{N_n} \rho_0 g(z_K - Z_K)\bar{V}_K\right) = \left(\frac{\partial}{\partial x}, \frac{\partial}{\partial y}, \frac{\partial}{\partial z}\right)_P \left(\sum_{K=1}^{N_n} \rho_0 g(z_K - Z_K)\bar{V}_K\right)$$

$$= (0, 0, \rho_0\, g) = \rho_0 g\bar{V}_P\, \delta_{i3},$$

$$(11.11)$$

where $j = 1$, 2, or 3 corresponding to the x, y, and z components of the term. Since Eq. 11.11 is true for every node, the "P" notation can be dropped and

$$\vec{\nabla}\left(\sum_{K=1}^{N_n} \rho_0 g(z_K - Z_K)\bar{V}_P\right) = \rho_0 g\, V_0\, \delta_{i3} \qquad (11.12)$$

using V_0 instead of $\bar{V}_P$.

The first term in Eq. 11.9 is the most difficult. The easiest way to evaluate this in Eq. 11.9 is to consider the nodes on a regular grid. For this section, the nodes will be initially located at X_{LMN} and finally located at x_{LMN}. The node located at X_{LMN} will be located at (L ΔX, M ΔY, N ΔZ). To help with the notation, define

$$x = x_1 \quad X = X_1$$
$$y = x_2 \quad Y = X_2 \qquad (11.13)$$
$$z = x_3 \quad Z = X_3$$

and express any specific node by adding a subscript "L M N", where L, M, and N run from 1 to the number of nodes in the x, y, and z directions, respectively. The first term in Equation 11.9 for the node at (L M N) is then

$$\vec{\nabla}_K\left(\sum_{I=1}^{N_v} \epsilon_I V_I\right) = \vec{\nabla}_{LMN}\left(\sum_{I=1}^{N_v} \epsilon_I V_I\right)$$

$$= \left(\frac{\partial\left(\sum_{I=1}^{N_v} \epsilon_I V_I\right)}{\partial x_{LMN}}, \frac{\partial\left(\sum_{I=1}^{N_v} \epsilon_I V_I\right)}{\partial y_{LMN}}, \frac{\partial\left(\sum_{I=1}^{N_v} \epsilon_I V_I\right)}{\partial z_{LMN}}\right). \quad (11.14)$$

Let us pause to consider the notation. We are taking the gradient of the node designated as L M N. For example, the gradient of the node that is the second node in the x direction, the fifth node in the y direction, and the tenth node in the z direction would be $\vec{\nabla}_{2,5,10}$.

First concentrate on the x component of the gradient in Eq. 11.14. Using the chain rule and that ϵ_I is a function of $\frac{\partial x_i}{\partial X_j}$, we get

$$\frac{\partial\left(\sum_{I=1}^{N_v}\epsilon_I V_I\right)}{\partial x_{LMN}} = \frac{\partial\left(\sum_{I=1}^{N_v}\epsilon_I V_I\right)}{\partial\left(\partial x_i/\partial X_j\right)_{IJK}} \frac{\partial\left(\partial x_i/\partial X_j\right)_{IJK}}{\partial x_{LMN}}, \tag{11.15}$$

where I, J, and K are summed from 1 to the number of nodes in the x, y, and z directions, respectively, and i and j are summed from 1 to 3 (x_i corresponding to the x component, x_2 corresponding to the y component, and x_3 corresponding to the z component). The sum over I, J, K, and i and j are not expanded because, for example, a $10 \times 10 \times 10$ node system would have 9000 terms!

Note that the partial derivatives, $\left(\frac{\partial x_i}{\partial X_j}\right)_{IJK}$, can be expressed as follows:

$$\frac{\partial x_1}{\partial X_1}\bigg|_{IJK} = \frac{\partial x}{\partial X}\bigg|_{IJK} = \left(\lim_{X_{IJK}\to X_{I-1JK}} \frac{x_{IJK} - x_{I-1JK}}{X_{IJK} - X_{I-1JK}}\right) = \left(\lim_{\Delta X\to 0} \frac{x_{IJK} - x_{I-1JK}}{\Delta X}\right)$$

$$\frac{\partial x_1}{\partial X_2}\bigg|_{IJK} = \frac{\partial x}{\partial Y}\bigg|_{IJK} = \left(\lim_{Y_{IJK}\to Y_{I-1JK}} \frac{x_{IJK} - x_{IJ-1K}}{Y_{IJK} - Y_{IJ-1K}}\right) = \left(\lim_{\Delta Y\to 0} \frac{x_{IJK} - x_{IJ-1K}}{\Delta Y}\right)$$

$$\frac{\partial x_1}{\partial X_3}\bigg|_{IJK} = \frac{\partial x}{\partial Z}\bigg|_{IJK} = \left(\lim_{Z_{IJK}\to Z_{IJK-1}} \frac{x_{IJK} - x_{IJK-1}}{Z_{IJK} - Z_{IJK-1}}\right) = \left(\lim_{\Delta Z\to 0} \frac{x_{IJK} - x_{IJK-1}}{\Delta Z}\right)$$

$$\frac{\partial x_2}{\partial X_1}\bigg|_{IJK} = \frac{\partial y}{\partial X}\bigg|_{IJK} = \left(\lim_{X_{IJK}\to X_{I-1JK}} \frac{y_{IJK} - y_{I-1JK}}{X_{IJK} - X_{I-1JK}}\right) = \left(\lim_{\Delta X\to 0} \frac{y_{IJK} - y_{I-1JK}}{\Delta X}\right)$$

$$\frac{\partial x_2}{\partial X_2}\bigg|_{IJK} = \frac{\partial y}{\partial Y}\bigg|_{IJK} = \left(\lim_{Y_{IJK}\to Y_{IJ-1K}} \frac{y_{IJK} - y_{IJ-1K}}{Y_{IJK} - Y_{IJ-1K}}\right) = \left(\lim_{\Delta Y\to 0} \frac{y_{IJK} - y_{IJ-1K}}{\Delta Y}\right)$$

$$\frac{\partial x_2}{\partial X_3}\bigg|_{IJK} = \frac{\partial y}{\partial Z}\bigg|_{IJK} = \left(\lim_{Z_{IJK}\to Z_{IJK-1}} \frac{y_{IJK} - y_{IJK-1}}{Z_{IJK} - Z_{IJK-1}}\right) = \left(\lim_{\Delta Z\to 0} \frac{z_{IJK} - z_{IJK-1}}{\Delta Z}\right)$$

$$\frac{\partial x_3}{\partial X_1}\bigg|_{IJK} = \frac{\partial z}{\partial X}\bigg|_{IJK} = \left(\lim_{X_{IJK}\to X_{I-1JK}} \frac{z_{IJK} - z_{I-1JK}}{X_{IJK} - X_{I-1JK}}\right) = \left(\lim_{\Delta X\to 0} \frac{z_{IJK} - z_{I-1JK}}{\Delta X}\right)$$

$$\frac{\partial x_3}{\partial X_2}\bigg|_{IJK} = \frac{\partial z}{\partial Y}\bigg|_{IJK} = \left(\lim_{Y_{IJK}\to Y_{IJ-1K}} \frac{z_{IJK} - z_{IJ-1K}}{Y_{IJK} - Y_{IJ-1K}}\right) = \left(\lim_{\Delta Y\to 0} \frac{z_{IJK} - z_{IJ-1K}}{\Delta Y}\right)$$

$$\frac{\partial x_3}{\partial X_3}\bigg|_{IJK} = \frac{\partial z}{\partial Z}\bigg|_{IJK} = \left(\lim_{Z_{IJK}\to Z_{IJK-1}} \frac{z_{IJK} - z_{IJK-1}}{Z_{IJK} - Z_{IJK-1}}\right) = \left(\lim_{\Delta Z\to 0} \frac{z_{IJK} - z_{IJK-1}}{\Delta Z}\right). \tag{11.16}$$

Notice that x_{IJK} appears in only the $\frac{\partial x_1}{\partial X_1}\big|_{IJK}$, $\frac{\partial x_1}{\partial X_1}\big|_{I+1JK}$, $\frac{\partial x_1}{\partial X_2}\big|_{IJK}$, $\frac{\partial x_1}{\partial X_2}\big|_{IJ+1K}$, $\frac{\partial x_1}{\partial X_3}\big|_{IJK}$, and $\frac{\partial x_1}{\partial X_3}\big|_{IJK+1}$ terms. As a result Eq. 11.15 reduces to

$$
\frac{\partial\left(\sum_{I=1}^{N_v}\epsilon_I V_I\right)}{\partial x_{LMN}} = \frac{\partial\left(\sum_{I=1}^{N_v}\epsilon_I V_I\right)}{\partial\left(\partial x/\partial X_j\right)_{LMN}}\frac{\partial\left(\partial x/\partial X_j\right)_{LMN}}{\partial x_{LMN}}
$$

$$
+\frac{\partial\left(\sum_{I=1}^{N_v}\epsilon_I V_I\right)}{\partial\left(\partial x/\partial X_j\right)_{L+1MN}}\frac{\partial\left(\partial x/\partial X_j\right)_{L+1MN}}{\partial x_{LMN}}
$$

$$
+\frac{\partial\left(\sum_{m=1}^{N_v}\epsilon_I V_I\right)}{\partial\left(\partial x/\partial X_j\right)_{LM+1N}}\frac{\partial\left(\partial x/\partial X_j\right)_{LM+1N}}{\partial x_{LMN}} \tag{11.17}
$$

$$
+\frac{\partial\left(\sum_{m=1}^{N_v}\epsilon_I V_I\right)}{\partial\left(\partial x/\partial X_j\right)_{LMN+1}}\frac{\partial\left(\partial x/\partial X_j\right)_{LMN+1}}{\partial x_{LMN}}
$$

with no sum over L, M, and N, but j is still summed from 1 to 3, so that Eq. 11.17 has 12 terms. All of the other terms are zero. Now evaluate the partial derivatives, $\frac{\partial\left(\partial x_i/\partial X_j\right)_{IJK}}{\partial x_{LMN}}$, using Eq. 11.16 to give X for X_1, Y for X_2, and Z for X_3 in the following:

$$
\frac{\partial\left(\partial x/\partial X_1\right)_{LMN}}{\partial x_{LMN}} = \frac{\partial\left(\partial x/\partial X\right)_{LMN}}{\partial x_{LMN}} = \left(\lim_{\Delta X\to 0}\frac{1}{\Delta X}\right)
$$

$$
\frac{\partial\left(\partial x/\partial X_1\right)_{L+1MN}}{\partial x_{LMN}} = \frac{\partial\left(\partial x/\partial X\right)_{L+1MN}}{\partial x_{LMN}} = \left(\lim_{\Delta X\to 0}\frac{-1}{\Delta X}\right)
$$

$$
\frac{\partial\left(\partial x/\partial X_2\right)_{LMN}}{\partial x_{LMN}} = \frac{\partial\left(\partial x/\partial Y\right)_{LMN}}{\partial x_{LMN}} = \left(\lim_{\Delta Y\to 0}\frac{1}{\Delta Y}\right)
$$

$$
\frac{\partial\left(\partial x/\partial X_2\right)_{LM+1N}}{\partial x_{LMN}} = \frac{\partial\left(\partial x/\partial Y\right)_{LM+1N}}{\partial x_{LMN}} = \left(\lim_{\Delta Y\to 0}\frac{-1}{\Delta Y}\right) \tag{11.18}
$$

$$
\frac{\partial\left(\partial x/\partial X_3\right)_{LMN}}{\partial x_{LMN}} = \frac{\partial\left(\partial x/\partial Z\right)_{LMN}}{\partial x_{LMN}} = \left(\lim_{\Delta Z\to 0}\frac{1}{\Delta Z}\right)
$$

$$
\frac{\partial\left(\partial x/\partial X_3\right)_{LMN+1}}{\partial x_{LMN}} = \frac{\partial\left(\partial x/\partial Z\right)_{LMN+1}}{\partial x_{LMN}} = \left(\lim_{\Delta Z\to 0}\frac{-1}{\Delta Z}\right)
$$

All other terms in Eq. 11.17 are zero, because these terms contain derivatives like

$$
\frac{\partial\left(\partial x/\partial X_1\right)_{LMN+1}}{\partial x_{LMN}} = \frac{\partial}{\partial x_{LMN}}\left(\lim_{\Delta X\to 0}\frac{x_{L+1MN+1} - x_{LMN+1}}{\Delta X}\right),
$$

$$
\frac{\partial\left(\partial x/\partial X_1\right)_{LM+1N}}{\partial x_{LMN}} = \frac{\partial}{\partial x_{LMN}}\left(\lim_{\Delta X\to 0}\frac{x_{L+1M+1N} - x_{LM+1N}}{\Delta X}\right), \tag{11.19}
$$

etc.

because these terms have no x_{LMN} in them.

    The four terms in Eq. 11.17 are then

$$\frac{\partial\left(\sum_{I=1}^{N_v}\epsilon_I V_I\right)}{\partial(\partial x/\partial X_j)_{LMN}}\frac{\partial(\partial x/\partial X_j)_{LMN}}{\partial x_{LMN}} = \frac{\partial\left(\sum_{I=1}^{N_v}\epsilon_I V_I\right)}{\partial(\partial x/\partial X_1)_{LMN}}\left(\lim_{\Delta X\to 0}\frac{1}{\Delta X}\right)$$

$$+\frac{\partial\left(\sum_{I=1}^{N_v}\epsilon_I V_I\right)}{\partial(\partial x/\partial X_2)_{LMN}}\left(\lim_{\Delta Y\to 0}\frac{1}{\Delta Y}\right)$$

$$+\frac{\partial\left(\sum_{I=1}^{N_v}\epsilon_I V_I\right)}{\partial(\partial x/\partial X_3)_{LMN}}\left(\lim_{\Delta Z\to 0}\frac{1}{\Delta Z}\right)$$

$$\frac{\partial\left(\sum_{I=1}^{N_v}\epsilon_I V_I\right)}{\partial(\partial x/\partial X_j)_{L+1MN}}\frac{\partial(\partial x/\partial X_j)_{L+1MN}}{\partial x_{LMN}} = \frac{\partial\left(\sum_{I=1}^{N_v}\epsilon_I V_I\right)}{\partial(\partial x/\partial X_1)_{L+1MN}}$$

$$\times\left(\lim_{\Delta X\to 0}\frac{-1}{\Delta X}\right), \qquad (11.20)$$

$$\frac{\partial\left(\sum_{I=1}^{N_v}\epsilon_I V_I\right)}{\partial(\partial x/\partial X_j)_{LM+1N}}\frac{\partial(\partial x/\partial X_j)_{LM+1N}}{\partial x_{LMN}} = \frac{\partial\left(\sum_{I=1}^{N_v}\epsilon_I V_I\right)}{\partial(\partial x/\partial X_2)_{LM+1N}}\left(\lim_{\Delta Y\to 0}\frac{-1}{\Delta Y}\right),$$

$$\frac{\partial\left(\sum_{I=1}^{N_v}\epsilon_I V_I\right)}{\partial(\partial x/\partial X_j)_{LMN+1}}\frac{\partial(\partial x/\partial X_j)_{LMN+1}}{\partial x_{LMN}} = \frac{\partial\left(\sum_{I=1}^{N_v}\epsilon_I V_I\right)}{\partial(\partial x/\partial X_3)_{LMN+1}}\left(\lim_{\Delta Z\to 0}\frac{-1}{\Delta Z}\right).$$

The sum over j in Eq. 11.17 now becomes

$$\frac{\partial\left(\sum_{I=1}^{N_v}\epsilon_I V_I\right)}{\partial x_{LMN}} = \frac{\partial\left(\sum_{I=1}^{N_v}\epsilon_I V_I\right)}{\partial(\partial x/\partial X_1)_{LMN}}\left(\lim_{\Delta X\to 0}\frac{1}{\Delta X}\right)$$

$$+\frac{\partial\left(\sum_{I=1}^{N_v}\epsilon_I V_I\right)}{\partial(\partial x/\partial X_1)_{L+1MN}}\left(\lim_{\Delta X\to 0}\frac{-1}{\Delta X}\right)$$

$$+\frac{\partial\left(\sum_{I=1}^{N_v}\epsilon_I V_I\right)}{\partial(\partial x/\partial X_2)_{LMN}}\left(\lim_{\Delta Y\to 0}\frac{1}{\Delta Y}\right)$$

$$+\frac{\partial\left(\sum_{I=1}^{N_v}\epsilon_I V_I\right)}{\partial(\partial x/\partial X_2)_{LM+1N}}\left(\lim_{\Delta Y\to 0}\frac{-1}{\Delta Y}\right) \qquad (11.21)$$

$$+\frac{\partial\left(\sum_{I=1}^{N_v}\epsilon_I V_I\right)}{\partial(\partial x/\partial X_3)_{LMN}}\left(\lim_{\Delta Z\to 0}\frac{1}{\Delta Z}\right)$$

$$+\frac{\partial\left(\sum_{I=1}^{N_v}\epsilon_I V_I\right)}{\partial(\partial x/\partial X_3)_{LMN+1}}\left(\lim_{\Delta Z\to 0}\frac{-1}{\Delta Z}\right).$$

Note that the derivative of $\left(\sum_{I=1}^{N_v}\epsilon_I V_I\right)$ with respect to $(\partial x/\partial X_1)_{LMN}$ is only nonzero for the ϵ_I terms that correspond to the node number, L M N. As a result,

$$\frac{\partial\left(\sum_{I=1}^{N_v}\epsilon_I V_I\right)}{\partial(\partial x/\partial X_1)_{LMN}} = \frac{\partial\epsilon_{LMN}V_{LMN}}{\partial(\partial x/\partial X_1)_{LMN}}. \tag{11.22}$$

Using Eq. 11.22 and rearranging Eq. 11.21, we have

$$\frac{\partial\left(\sum_{I=1}^{N_v}\epsilon_I V_I\right)}{\partial x_{LMN}} = -\left(\begin{array}{c} \lim_{\Delta X \to 0} \dfrac{\dfrac{\partial\left((\epsilon V_0)_{L+1MN}\right)}{\partial(\partial x/\partial X_1)_{L+1MN}} - \dfrac{\partial\left((\epsilon V_0)_{LMN}\right)}{\partial(\partial x/\partial X_1)_{LMN}}}{\Delta X} \\[2em] + \lim_{\Delta Y \to 0} \dfrac{\dfrac{\partial\left((\epsilon V_0)_{LM+1N}\right)}{\partial(\partial x/\partial X_2)_{LM+1N}} - \dfrac{\partial\left((\epsilon V_0)_{LMN}\right)}{\partial(\partial x/\partial X_2)_{LMN}}}{\Delta Y} \\[2em] + \lim_{\Delta Z \to 0} \dfrac{\dfrac{\partial\left((\epsilon V_0)_{LMN+1}\right)}{\partial(\partial x/\partial X_3)_{LMN+1}} - \dfrac{\partial\left((\epsilon V_0)_{LMN}\right)}{\partial(\partial x/\partial X_3)_{LMN}}}{\Delta Z} \end{array}\right) \tag{11.23}$$

or

$$\frac{\partial\left(\sum_{I=1}^{N_v}\epsilon_I V_I\right)}{\partial x_{LMN}} = -\left(\begin{array}{c}\dfrac{\partial}{\partial X}\left(\dfrac{\partial\left((\epsilon V_0)_{LMN}\right)}{\partial(\partial x/\partial X_1)_{LMN}}\right) + \dfrac{\partial}{\partial Y}\left(\dfrac{\partial\left((\epsilon V_0)_{LMN}\right)}{\partial(\partial x/\partial X_2)_{LMN}}\right) \\[1.5em] + \dfrac{\partial}{\partial Z}\left(\dfrac{\partial\left((\epsilon V_0)_{LMN}\right)}{\partial(\partial x/\partial X_3)_{LMN}}\right)\end{array}\right) \tag{11.24}$$

and finally

$$\frac{\partial\left(\sum_{I=1}^{N_v}\epsilon_I V_I\right)}{\partial x_{LMN}} = -\left(\frac{\partial}{\partial X_j}\frac{\partial(\epsilon V_0)_{LMN}}{\partial(\partial x/\partial X_j)_{LMN}}\right) \tag{11.25}$$

summed over $j = 1$, 2, and 3.

Since this applies to every node, L M N can be dropped, and we can just write

$$\frac{\partial\left(\sum_{I=1}^{N_v}\epsilon_I V_I\right)}{\partial x} = -\frac{\partial}{\partial X_j}\frac{\partial(\epsilon V_0)}{\partial\left(\frac{\partial x}{\partial X_j}\right)}. \tag{11.26}$$

This derivation can be repeated for $\dfrac{\partial\left(\sum_{I=1}^{N_v}\epsilon_I V_I\right)}{\partial y}$ and $\dfrac{\partial\left(\sum_{I=1}^{N_v}\epsilon_I V_I\right)}{\partial z}$ giving for the first term in Eq. 11.9,

$$\frac{\partial\left(\sum_{I=1}^{N_v}\epsilon_I V_I\right)}{\partial x_j} = -\frac{\partial}{\partial X_j}\frac{\partial(\epsilon V_0)}{\partial\left(\frac{\partial x_i}{\partial X_j}\right)}, \tag{11.27}$$

where j is summed over 1, 2, and 3, but i is not summed over and identifies three equations: one in x, one in y, and one in z.

Putting Eq. 11.9 back together, we get for Eq. 11.3,

$$m\ a_i = -\frac{\partial}{\partial x_i}E_{tot} = +\frac{\partial}{\partial X_j}\frac{\partial(\epsilon V_0)}{\partial(\partial x_i/\partial X_j)} - \rho_0\ g\ V_0\ \delta_{i3}, \tag{11.28}$$

with $\frac{\vec{F}}{A_0}$ supplying the $\frac{\partial\epsilon}{\partial(\partial x_i/\partial X_j)}$ values for the boundary nodes.

Equation 11.28 is sufficient for describing discrete nodes, but not for a continuous equation, because both sides of Eq. 11.28 approach zero as the separation of the nodes approaches zero. To get a continuous equation, divide both sides of Eq. 11.28 by the initial volume of each node, V_0. V_0 is a constant with respect to X_j and $\partial x_i/\partial X_j$ so that it passes through these derivatives. Equation 11.28 becomes

$$\rho_0\ a_i = +\frac{\partial}{\partial x_j}\left(\frac{\partial\epsilon}{\partial(\partial x_i/\partial x_j)}\right) - \rho_0\ g\ \delta_{i3}, \tag{11.29}$$

where

$$\rho_0 = \lim_{\substack{V_0 \to 0 \\ N_v \to \infty}}\frac{m}{V_0}, \tag{11.30}$$

and this is the same as Eq. 4.45 so that the derivation is complete.

Force on Fixed Nodes

Equation 11.1 can also be used in simulations to find the constraint force on any fixed node or node component. The sum of the forces on a fixed node is zero, so

$$-\vec{\nabla}E_{tot} = F_{tot} = \Sigma F_m + \Sigma F_g + \Sigma F_B = 0, \tag{11.31}$$

where

F_m = force from the material on the fixed node
F_g = force from gravity on the fixed node
F_B = boundary force on the fixed node.

Equation 11.31 can be solved for the total boundary force on the fixed node,

$$\Sigma F_B = -\left(\Sigma F_m + \Sigma F_g\right). \tag{11.32}$$

In a simulation, no boundary forces are applied to fixed nodes, so that $\Sigma F_B = 0$, and Equation 11.31 gives

$$-\vec{\nabla} E_{tot} = \Sigma F_m + \Sigma F_g \tag{11.33}$$

or

$$\vec{\nabla} E_{tot} = -\left(\Sigma F_m + \Sigma F_g\right). \tag{11.34}$$

Now let's ask the question, what boundary force would be necessary to hold the node fixed, if the node were to be allowed to move? In that case F_B would not be zero in Eq. 11.31, and we could solve Eq. 11.31 for the needed boundary force,

$$\Sigma F_B = -\left(\Sigma F_m + \Sigma F_g\right), \tag{11.35}$$

which by Eq. 11.34 is just the gradient of the total energy calculated in the fixed node simulation. Thus for fixed nodes in simulations, the boundary forces on the fixed node are

$$\Sigma F_B = \vec{\nabla} E_{tot}. \tag{11.36}$$

Equation 11.36 shows that although we do not compute the gradient of E_{tot} to find the final position of fixed nodes, the gradient of E_{tot} can be used to find the force required to hold fixed nodes in place.

Numerical Simulation

For a numerical simulation example of time-dependent simulation, I have chosen to use **NDSolve** in Mathematica. **NDSolve** will solve coupled partial differential equations. The input consists of a list of the differential equations to be solved, the functions to be solved for, and the variable range for which solutions are desired. For this case, the form is as follows:

NDSolve[{equations of motion for each node which can move, nodes which can move in time, time range to solve over}]

The equations of motion for each node that can move are Eq. 11.3, where the gradient is taken over the component of every node that is not fixed. This gives N equations, where N = number of node components which can change (or are not fixed).

For this numerical simulation, I have chosen to first deform a cylinder and then release it to follow its time-dependent motion. To accomplish this, I first run the cylindrical compression simulation, described in section "Cylindrical compression test" of Chap. 8. I use the final node positions of this simulation to be the starting place of the time-dependent solution. Thus, the initial position of each node in the timed simulation is the final position of each node after a quasi-static deformation simulation has been run. The final location of each node that was allowed to move in the quasi-static simulation is stored in **varsi0**. The form of **varsi0** is

$$\mathbf{varsi0} = \{\{x[1], 2\}, \{y[1], 3.1\}, \dots \{x[3], 4.\} \dots\}, \tag{11.37}$$

where

x[1] = x coordinate of node 1, which has a value of 2 at the end of the quasi-static deformation
y[1] = y coordinate of node 1, which has a value of 3.1 at the end of the quasi-static deformation
. . .
x[3] = x coordinate of node 3, which has a value of 2 at the end of the quasi-static deformation
. . .
etc.

For the time-dependent simulation, E_{tot} in Eq. 11.3 is

$$E_{tot} = (\mathbf{etot} - \mathbf{bcst}) + \mathbf{gravity}, \tag{11.38}$$

where

etot = energy of deformation from quasi-static deformation expressed as a function of **x[i],y[i],z[i]** for i = 1 to N
bcst = boundary force contributions to E_{tot} (set to 0 for these simulations)
gravity = energy due to gravity (set to 0 for these simulations).

(Note that Eq. 11.38 is the same form as used in the quasi-static deformation.) Then $\overrightarrow{\nabla} E_{tot}$ is

$$- \mathbf{D}[(\mathbf{etot} - \mathbf{bcst}) + \mathbf{gravity}, \mathbf{varsi0}[[i, 1]]] \tag{11.39}$$

for $i = 1$ Eq. 11.39 is $\frac{\partial}{\partial x[1]} (E_{tot})$, for i=2, $\frac{\partial}{\partial y[1]} (E_{tot})$, etc.

The result expresses each moveable component as **x[I]**, **y[I]**, or **z[I]**. But we wish these to be functions of time, so a substitution **varsi0[[i,1]]→varsi0[[i,1]][t]** is made after the derivative is taken.

Up to now, we have not discussed friction. To add interest to this simulation, a frictional force, F_{friction}, is added to each node,

$$F_{\text{friction}} = -\mu \ \text{v}, \tag{11.40}$$

where v is the velocity of the node and μ is the coefficient of friction.

The component of friction is expressed in the simulation as $-\mu$ **varsi0[[i,1]]'[t]**. The left-hand side of Eq. 11.3 is a_i, where

$$\text{m} = \rho_0 \ V_0, \tag{11.41}$$

where $V_0 =$ initial volume associated with the node. For this simulation I have chosen, $\rho_0 = 100$. The mass associated with each node is calculated as $\rho_0 V_0$. (In all the computer entries, ρ is ρ_0.) A list of point volumes are calculated in **pv** using **pointVols[pts,polys,k]** for the kth mass. Since there are equations for each free component of each node, the volume associated with the ith equation is found from the **varsi[[i,1,1]]** term in the variable list, so that

$$\text{m} = \rho \ \textbf{pv}[[\textbf{varsi}[[\textbf{i, 1, 1, 1}]]]]. \tag{11.42}$$

The acceleration of each node, a_i, is

$$a_i = \frac{\partial^2}{\partial t^2} \ \textbf{varsi0}[[\textbf{i, 1}]] = \textbf{varsi0}[[\textbf{i, 1}]]''[\textbf{t}], \tag{11.43}$$

so that **varsi0[[i,1]]''[t]** is explicitly a function of time.

A table of

$$\rho_0 \ V_0 \ \frac{\partial^2 x_i}{\partial t^2} = -\frac{\partial}{\partial x_i} E_{\text{tot}} - F_{\text{friction}} \tag{11.44}$$

for all the components for all the nodes that are not fixed in the simulation makes up the list of differential equations to be solved.

We need boundary conditions for time. For each node there are two: One stating the initial position of the node (t = 0),

$$\textbf{varsi0}[[\textbf{i, 1}]][0] = = \textbf{varsi0}[[\textbf{i, 2}]], \tag{11.45}$$

and the second stating that the initial velocity of each node is zero, i.e.,

$$\textbf{varsi0}[[\textbf{i, 1}]]'[\textbf{0}] = = \textbf{0}. \tag{11.46}$$

A list of all the nodes to be solved for is made with

$$\textbf{Table}[\ \textbf{varsi0}[[\textbf{i, 1}]],\ \{\textbf{i, 1, Length}[\textbf{varsi0}]\}], \tag{11.47}$$

and the time range over which the solution is desired is expressed as

$$\{\textbf{t, 0, tmax}\}. \tag{11.48}$$

Lastly in the NDSolve, the method is specified as

$$\textbf{Method} \!\!-\!\!> \big\{ {}^{\text{“}}\textbf{EquationSimplification}{}^{\text{”}} \!\!-\!\!> {}^{\text{“}}\textbf{Residual}{}^{\text{”}}\big\}. \tag{11.49}$$

The output of NDSolve is an interpolating function for each variable. Any variable can be converted into a numerical value using the form

$$\textbf{x}[\textbf{1}][\textbf{t}]/.\textbf{s}/.\textbf{t} \rightarrow \textbf{2.7}, \tag{11.50}$$

where "s" is the output of **NDSolve** and the time chosen to express the x component of node 1 is 2.7 in the same units as **tmax**.

The simulation is run for a **tmax** of 100. The output of the first 18 time steps in increments of 2 is shown in Fig. 11.2. Note the oscillation of the cylinder. The oscillation begins at time = 0 as a result of the initial quasi-static simulation. The compression force is removed, and the cylinder begins to rise to take on its original shape, which is shown in the deeper blue color in Fig. 11.2. The cylinder returns to its original shape about time step 6 and then continues to rise because the nodes in the cylinder are traveling with a velocity at time step 6. As a result, the cylinder overshoots its initial configuration and continues rising until the restorative internal forces stop the rise. Next the cylinder begins to compress back toward its original shape. In subsequent time steps (not shown in Fig. 11.2), the cylinder overshoots its equilibrium position and continues down to its original position and then begins to be drawn back upwards. This oscillation would continue forever if there was no frictional force to stop the oscillations.

Figure 11.2 is a bit messy to view, so consider a single node on the top of the cylinder (node 50). Figure 11.3 shows a plot of node 50's z component as a function of time for 100 time steps.

The time-dependent simulation can be repeated for different amounts of friction. More friction tends to stop the oscillations faster. Figures 11.4, 11.5 and 11.6 show examples of increasing the friction, μ.

The last thing I will do with this simulation is to calculate the force on the bottom of the cylinder at the beginning of the time-dependent simulation. Equation 11.36 can be used to compute the boundary force on the bottom of the cylinder.

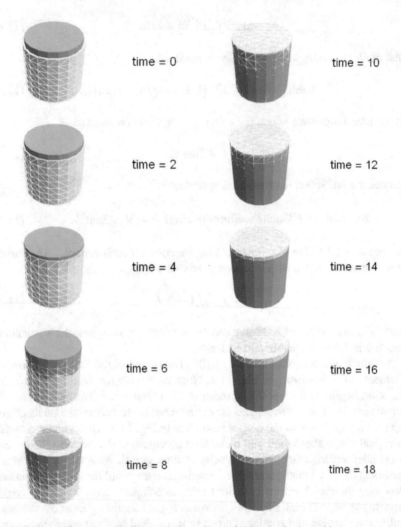

Fig. 11.2 Simulation of a cylinder oscillating without friction beginning with a compression of 10% at time = 0

Equation 11.36 could be used "as is", but for problems with many nodes, this is not efficient. E_{tot} is the sum of the energy contributed by all the tetrahedra in the material. But only those tetrahedra that contain the target node will produce a nonzero gradient of E_{tot}. As a result, it is more efficient in large simulations to isolate those tetrahedra which contain the target node and calculate the derivative of the energy for only those tetrahedra (i.e., replace E_{tot} = sum over all tetrahedra with the energy with E_{tot} = sum only over those tetrahedra that contain the target node).

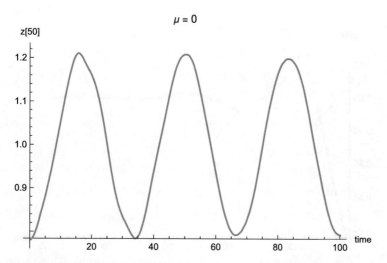

Fig. 11.3 Oscillation of node 50 on the top of the cylinder. Plotted is the z component versus time for a simulation with no friction ($\mu = 0$). Note that the oscillation varies about the original location of the top of the cylinder ($z = 1.0$)

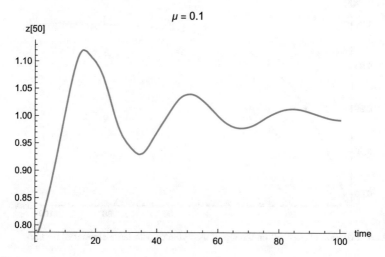

Fig. 11.4 Location of node 50 on the top of the cylinder with a small amount of friction

Figure 11.7 shows an example of the tetrahedra that will contribute to the energy of a single node at the bottom of the cylinder.

The result of calculating the force on all the bottom nodes is shown in Fig. 11.8.

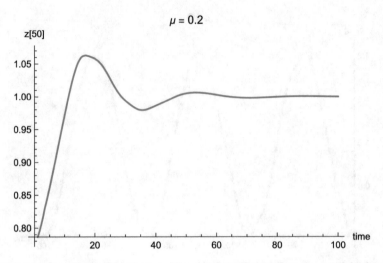

Fig. 11.5 Location of node 50 on the top of the cylinder with a medium amount of friction

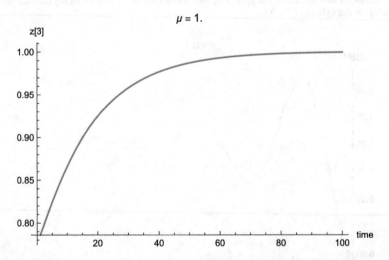

Fig. 11.6 Location of node 50 on the top of the cylinder with large amount of friction

Note that all the forces are in the $+z$ direction at the bottom of the cylinder. This is because in this simulation the cylinder is allowed to expand in the x and y directions without constraint, so there are no nonzero components of the constraint force in these directions on the bottom of the cylinder. Figure 11.8 shows that the constraint force on the nodes on the edge of the cylinder are less than the constraint forces in the interior of the cylinder. This may seem curious at first, but this difference disappears if we calculate the pressure – or force per unit area – for each node. When this is done, the pressure on the bottom of the cylinder is seen to be uniform (Fig. 11.9).

Fig. 11.7 Tetrahedra that participate with an energy change when a single target node is moved. There are 1680 tetrahedra, but only the 10 have shown change with the movement of the target node

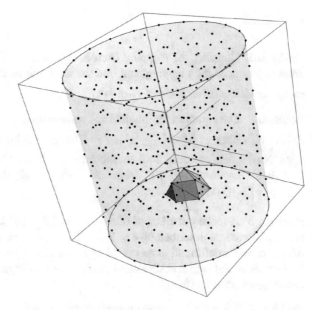

Fig. 11.8 Force on the nodes on the bottom of the cylinder after the quasi-static deformation

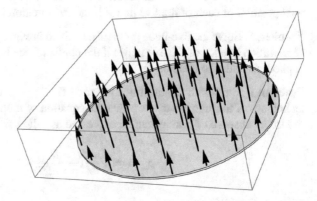

Fig. 11.9 Force per unit area on the bottom nodes of the cylinder after the quasi-static deformation

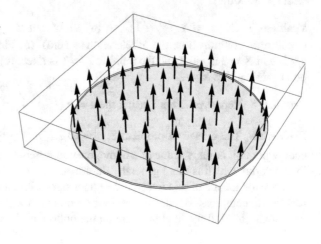

Problems

Problem 1 Consider the mapping from X to $x = 2/m^2\, X^3$. Assume a quasi-static process produced the mapping, and find the net force on a point at $X = 8$ m if the energy of deformation is $\epsilon = 4$ N-m $\left(\frac{\partial x}{\partial X}\right)^2$.

Problem 2 The equation for longitudinal waves (small displacements) in an elastic rod can be written as $\frac{\partial^2 \xi}{\partial t^2} = \frac{Y}{\rho}\frac{\partial^2 \xi}{\partial X^2}$, with Y = Young's modulus, ρ = density of the rod, and $\xi(X,t)$ = displacement from equilibrium (i.e., $\xi = x - X$). Use Eq. 11.3 to show that this equation follows from an energy of deformation defined as $\epsilon = 1/2Y\left(\frac{\partial x}{\partial X} - 1\right)^2$.

Problem 3 A rectangular bar of rubber, $\rho_m = 1.2 * 10^3$ kg/m^3, has an energy per unit initial volume described by $\epsilon = Y\,(dx/dX - 1)^3$ with $Y = 0.2$ N/m^2. The bar is stretched with a deformation described by $x = a\,X^2$ for X>0 and $a = 6/m$. The bar is then released. Find the acceleration of the point at $x = 24$ m when the bar is released. Ignore gravitational effects.

Problem 4 Solve the five-node problem described in Fig. 11.1 in this chapter for the final position of each of the five nodes if the deformation is quasi-static.

Problem 5 Solve the five-node problem described in Fig. 11.1 in this chapter for the acceleration of each of the five nodes if the applied force is an impulse force (i.e., is applied instantly) to node 5.

Problem 6 Start with the net force on node K, $F_K = -\, d\{E_{tot}\}/dx_K$, and use the techniques of this chapter to show that the equation of motion of a one-dimensional rod in the absence of gravity can be expressed as follows:

$$\rho_m\, a = +\frac{d}{dx}\left(\frac{d\epsilon}{(dx/dX)}\right).$$

with $\epsilon = f(dx/dX)$.

Problem 7 Use $\epsilon = Y\left(\frac{dx}{dX} - 1\right)^4$ to (a) show that $x = A\,X + B$ provides an equilibrium solution (i.e., the acceleration is zero). (b) Find A and B if the force applied at $X = 3$ is -0.2 and the point at $X = 1$ is fixed. (c) Find the location of the point originally at $X = 2$.

Problem 8 Repeat problem 7 using the $\epsilon = b\left(\left(\frac{dx}{dX}\right)^2 + \frac{2}{dx/dX} - 3\right)$, with $b = 0.2$.

Problem 9 The ϵ used in problem 8, $\epsilon = (0.2\ N/m^2)\left(\left(\frac{dx}{dX}\right)^2 + \frac{2}{dx/dX} - 3\right)$, is the energy per unit initial volume of a homogeneous, symmetric, isotropic, incompressible material, which is more generally described with $\epsilon = (0.2\ N/m^2)\,I_1$. Consider a compression a cube of this material 2 m on a side, without shear. (a) Find the force required to compress the cube to 1 m in the x direction with no forces on the y and z directions. (b) Find the final pressure on the bottom of the cube.

Chapter 12
Anisotropic Materials

So far, we have dealt with only isotropic materials. The basic equations of motion are the same for both anisotropic and isotropic materials (Eq. 4.45). All that needs to be changed for anisotropic materials is the representation of the energy, ϵ, of the material. For isotopic materials, we changed the general energy function, $\epsilon = f(\partial x_i/\partial X_j)$, to the more restrictive $\epsilon = f(I_i)$, where the I_i values were the invariants of the $\partial x_i/\partial X_j$ matrix. In that case, the I_i values did not change when the material was rotated both before and after a deformation. For anisotropic materials, we need to change ϵ to a function of $\partial x_i/\partial X_j$ that is invariant only for rotations after the deformation. To do this, we will introduce a different set of invariants. Three invariants were needed to express the energy for isotropic materials; six are needed for anisotropic materials.

In this chapter, we will first discuss a couple of example anisotropic materials. Next, we will show how to take measurements to define ϵ for an anisotropic material. To do this, a homogeneous sample is placed in a jig, and measurements are made similar to those for isotropic materials described in Chap. 10; only this time the jig must allow for six different deformations instead of only three. To use this in a general deformation of the material, $\mathcal{F}_m$, we also need to know the initial orientation of anisotropy in the material when it is placed in service. From this orientation, we can express any local material deformation expressed in fixed coordinates, $\mathcal{F}_m$, in terms of a deformation in jig coordinates, $\mathcal{F}_{jig}$. The column vectors of $\mathcal{F}_{jig}$ can then be used to calculate ϵ. As with the isotropic case, once we have ϵ as a function of $\partial x_i/\partial X_j$, we can calculate the forces and displacements of the material.

This chapter also contains four simulation examples of anisotropic "materials", two simple spring models and two more complex, many node examples. These examples will illustrate the steps in simulating the deformation of an anisotropic material. In these examples, jig experiments were not required to define ϵ because we

Supplementary Information The online version contains supplementary material available at [https://doi.org/10.1007/978-3-031-09157-5_12].

were able to infer ϵ from the known properties of the materials causing the anisotropy, i.e., the springs and their geometry.

Anisotropic Materials

As an example of natural anisotropic materials, consider wood (Fig. 12.1). The obvious grains in wood grain may make the wood anisotropic. For example, wood may twist or shrink unevenly when it dries as a result of different properties of the wood in the different layers. To simulate anisotropic materials, it is necessary to isolate the anisotropy and create a coordinate system in which to define the anisotropy. We will call this the jig coordinate system. In Fig. 12.1, two jig coordinate systems are superimposed on the wood and are oriented in different directions. In each case the x-y plane of the jig coordinate system is parallel to the wood grain. The third coordinate system, placed off the wood, represents the fixed coordinate system to be used to calculate the deformation of the wood as a whole.

The wood in Fig. 12.1 is not only anisotropic, it is also inhomogeneous because the wood grain is different in different parts of the wood. As an example of a material that is anisotropic but homogeneous, consider a cube of the same Dragon Skin material experimented with in Chap. 10, but with stiff, linear elastic fibers placed parallel to the x-axis. These fibers change the elastic properties of the material along the x-axis but have little effect on the elastic properties along the y- and z-axes. The material is homogenous in that only one jig coordinate system is needed to describe the anisotropy throughout the material. Figure 12.2 shows the jig coordinate system oriented so that the x-axis parallels the fibers.

Fig. 12.1 One fixed coordinate system off the wood and two different jig coordinate systems defined along grain direction in a piece of wood

Fig. 12.2 Cartoon of Dragon Skin with elastic fibers parallel to the x-axis. Only one coordinate system is needed to capture the anisotropy

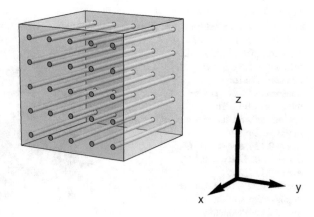

Fig. 12.3 Imaginary jig showing the degrees of freedom allowed. The jig is designed so that (0,0,0) is fixed. The (1,0,0) corner can only be deformed as $(1 + dxx, 0, 0)$. The (0,1,0) corner can only be deformed as $(dyx, 1 + dyy, 0)$. The (0,0,1) corner can only be deformed as $(dzx, dzy, 1 + dzz)$

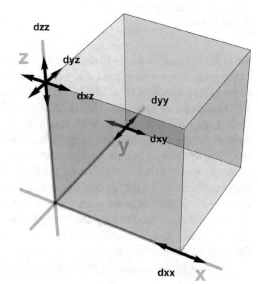

Measuring the Energy of Deformation of an Anisotropic Material

Once the anisotropy of a material has been identified and a portion of the material is placed in a jig, ϵ needs to be measured. We can do this by making jig measurements on a homogeneous sample of the material – preferably a cube. Three edges of the cube that intersect in a single point will define the jig coordinate system (Fig. 12.3).

Next, we need to describe how to experimentally find the energy function, ϵ. Orient the sides of our cubic jig so that they are aligned with the anisotropy (e.g., Fig. 12.2). Fix the bottom left corner (0,0,0) of the sample so it will not move. Allow

Fig. 12.4 Example of the cube inside the jig before and after deformation. The more transparent cube is the material in the original position. The less transparent "cube" is after the deformation. The applied deformation shown in the example is $dxx = 0.124$, $dyy = -0.165$, $dzz = -0085$, $dxy = 0.152$, $dxz = 0.137$, and $dyz = -0.119$. Note that the opposite faces of the material are identically deformed and parallel after the deformation

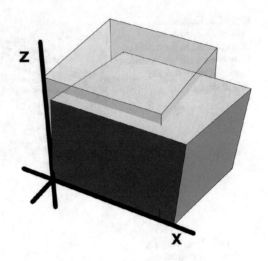

the (1,0,0) corner of the jig to move only in the x direction. Allow the (0,1,0) corner of the jig to move only in the x and y directions. Allow the (0,0,1) corner of the jig to move in any direction (Fig. 12.3). The jig is so designed that opposite parallel faces will always move in unison. Forces can be applied at four points on the jig, (0,0,0), (1,0,0), (0,1,0), and (0,0,1). With this arrangement, forces acting on the jig will deform the material homogeneously. We can deform the jig with movements in the dxx, dxy, dyy, dxz, dyz, and dzz directions as shown in Fig. 12.4.

Apply forces to the four points on the jig, (0,0,0), (1,0,0), (0,1,0), and (0,0,1). As the material deforms, the applied forces will do work on the material. The total work done by all the forces is the energy stored in the material. Many different deformations can be performed to map out the energy as a function of the displacements in a manner similar to that described in Chap. 10, but for anisotropic materials, ϵ will be a function of six variables instead of only three.

The four points on the jig before the deformation are

$$\begin{aligned}
\vec{B}_1 &= (0, 0, 0) \\
\vec{B}_2 &= (1, 0, 0) \\
\vec{B}_3 &= (0, 1, 0) \\
\vec{B}_4 &= (0, 0, 1).
\end{aligned} \tag{12.1}$$

After the deformation, these points map to the following points:

$\vec{b}_1 = (0, 0, 0)$ this point is fixed

$\vec{b}_2 = (1 + dxx, 0, 0)$ this point is allowed to move only in the x direction

$\vec{b}_3 = (dxy, 1 + dyy, 0)$ this point is allowed to move only in the x and y directions

$\vec{b}_4 = (dxz, dyz, 1 + dzz)$ this point is allowed to move in any direction.

$$(12.2)$$

From the location of these four points, we can construct the $\mathcal{F}_{\text{jig}}$ matrix that maps any point in the material to a new point in the material. We do this by noting that

$$\vec{b}_i = \mathcal{F}_{\text{jig}} \vec{B}_i \qquad (12.3)$$

for $i = 2, 3,$ and 4. Thus

$$\begin{pmatrix} 1 & + & dxx \\ 0 & & \\ 0 & & \end{pmatrix} = \begin{pmatrix} \mathcal{F}_{11} & \mathcal{F}_{12} & \mathcal{F}_{13} \\ \mathcal{F}_{21} & \mathcal{F}_{22} & \mathcal{F}_{23} \\ \mathcal{F}_{31} & \mathcal{F}_{32} & \mathcal{F}_{33} \end{pmatrix} \begin{pmatrix} 1 \\ 0 \\ 0 \end{pmatrix} \qquad (12.4)$$

$$\begin{pmatrix} dxy \\ 1 + dyy \\ 0 \end{pmatrix} = \begin{pmatrix} \mathcal{F}_{11} & \mathcal{F}_{12} & \mathcal{F}_{13} \\ \mathcal{F}_{21} & \mathcal{F}_{22} & \mathcal{F}_{23} \\ \mathcal{F}_{31} & \mathcal{F}_{32} & \mathcal{F}_{33} \end{pmatrix} \begin{pmatrix} 0 \\ 1 \\ 0 \end{pmatrix} \qquad (12.5)$$

$$\begin{pmatrix} dxz \\ dyz \\ dzz \end{pmatrix} = \begin{pmatrix} \mathcal{F}_{11} & \mathcal{F}_{12} & \mathcal{F}_{13} \\ \mathcal{F}_{21} & \mathcal{F}_{22} & \mathcal{F}_{23} \\ \mathcal{F}_{31} & \mathcal{F}_{32} & \mathcal{F}_{33} \end{pmatrix} \begin{pmatrix} 0 \\ 0 \\ 1 \end{pmatrix}. \qquad (12.6)$$

Solving Eqs. 12.4, 12.5, and 12.6 gives

$$\begin{aligned} \mathcal{F}_{11} &= 1 + dxx \\ \mathcal{F}_{12} &= dxy \\ \mathcal{F}_{22} &= 1 + dyy \\ \mathcal{F}_{13} &= dxz \\ \mathcal{F}_{23} &= dyz \\ \mathcal{F}_{33} &= 1 + dzz \end{aligned} \qquad (12.7)$$

with all the other elements of $\mathcal{F}_{\text{jig}}$ are zero.

Since the material is homogeneous, $\mathcal{F}_{\text{jig}}$ is the same for every point within the body. Thus, the displacements of the material correspond to an upper triangular matrix, $\mathcal{T}$, defined as

$$\mathcal{F}_{\text{jig}} = \begin{pmatrix} 1+dxx & dxy & dxz \\ 0 & 1+dyy & dyz \\ 0 & 0 & 1+dzz \end{pmatrix} = \begin{pmatrix} \mathcal{T}_{11} & \mathcal{T}_{12} & \mathcal{T}_{13} \\ 0 & \mathcal{T}_{22} & \mathcal{T}_{23} \\ 0 & 0 & \mathcal{T}_{33} \end{pmatrix} = \mathcal{T}. \quad (12.8)$$

It is the elements of $\mathcal{T}$ which will make up ϵ. By measuring the change in position of each of the four points, we have found $\mathcal{T}$. By measuring the work done by the applied forces to produce a particular $\mathcal{T}$, we have the energy corresponding to that $\mathcal{T}$. Performing many such measurements provides the data needed to fit a function $\epsilon(\mathcal{T})$.

How to Find the Jig Coordinates Within a Sample

Two examples will be used to show how to find the jig coordinates from the initial orientation of the anisotropy material within a sample. The fixed coordinates are the coordinates in which we will carry out simulations for applications. The jig coordinates are the coordinate system used to measure ϵ as a function of $\mathcal{T}$. The two examples illustrate how to use these two coordinate systems with anisotropic materials. In both cases, one vector needs to be projected onto another vector. This vector projection is described next.

Vector Projection

This section describes how to project one vector along another vector. Fig. 12.5 shows the projection of $\vec{a}$ along $\vec{e}$. The projection of $\vec{a}$ along $\vec{e}$ is a vector that is parallel to $\vec{e}$ (i.e., $\text{proj}_e\,\vec{a}$) having a length equal to the perpendicular projection of $\vec{a}$ onto $\vec{e}$. The figure also shows the vector $\vec{v}$ which is perpendicular to $\vec{e}$ so that we have Eq. 12.9.

$$\vec{a} = \text{proj}_e\,\vec{a} + \vec{v}. \quad (12.9)$$

The projection of $\vec{a}$ along $\vec{e}$ is carried out with the following formula:

$$\text{proj}_e\,\vec{a} \cdot = \frac{\vec{a} \overset{\circ}{\cdot} \vec{e}}{\left(\vec{e} \cdot \vec{e}\right)}\, \vec{e}. \quad (12.10)$$

Note that $\dfrac{\vec{e}}{\sqrt{\vec{e} \cdot \vec{e}}}$ is a unit vector along $\vec{e}$. The length of $\vec{a}$ along $\vec{e}$ is

Fig. 12.5 The projection of $\vec{a}$ along $\vec{e}$ is shown in red as $\mathrm{proj}_e\,\vec{a}$. $\vec{v}$ is perpendicular to $\vec{e}$ in the plane defined by $\vec{a}$ and $\vec{e}$

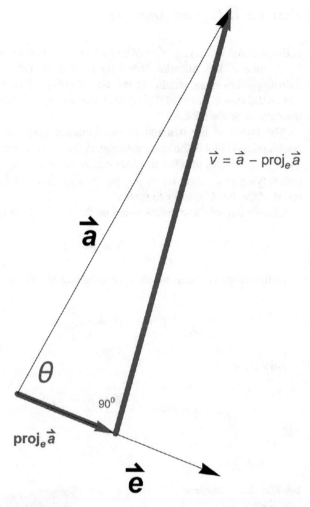

$$\vec{a} \cdot \left(\frac{\vec{e}}{\sqrt{\vec{e} \cdot \vec{e}}} \right) = |\,\vec{a}\,|\ \cos\theta, \qquad (12.11)$$

where θ is the angle between $\vec{a}$ and $\vec{e}$. The $\mathrm{proj}_e\,\vec{a}$ is the product of the unit vector along $\vec{e}$ and the magnitude of $\vec{a}$ along $\vec{e}$ and is given in Eq. 12.10.

Also note that the vector $\vec{v}$ defined as

$$\vec{v} = \vec{a}\,\mathrm{proj}_e\,\vec{a}. \qquad (12.12)$$

is perpendicular to the vector $\vec{e}$ and lies in the plane defined by $\vec{e}$ and $\vec{a}$. With these definitions we are prepared to define the orientation of the anisotropy in a material.

Example 1: Layered Anisotropy

If the material has layers of anisotropy like the wood example in Fig. 12.1, place a small cube of this material in the jig so that the cube is homogeneous and the anisotropic layers are parallel to the *x-y* axis (Fig. 12.6). Define the jig coordinates so that the axes intersect at (0,0,0), and with a proper choice of units, Eq. 12.1 define the corners of the cube.

The layers in the material to be simulated may not be parallel to the fixed coordinates. If that is the case, we need to find the rotation matrix, $\mathcal{R}$, which rotates the jig coordinates into the fixed coordinates. To find $\mathcal{R}$, choose three noncolinear points lying in a single layer, $\vec{p}_1$, $\vec{p}_2$, and $\vec{p}_3$ (Fig. 12.7). Express these points in terms of the fixed coordinate system.

Choose two of these points to define the *x*-axis of the jig coordinates, e.g.,

$$\vec{u} = \vec{p}_2 - \vec{p}_1. \tag{12.13}$$

Define $\widehat{B}_1$ to be a unit vector corresponding to the *x*-axis of the jig coordinates

$$\widehat{B}_1 = \frac{\vec{u}}{\sqrt{\vec{u}\overset{\circ}{}\vec{u}}}. \tag{12.14}$$

Now define

$$\vec{q} = \vec{p}_3 - \vec{p}_1 \tag{12.15}$$

and

Fig. 12.6 Small sample of layered anisotropy oriented in a jig so that the anisotropic layer is parallel to the *x-y* plane of the jig coordinate system

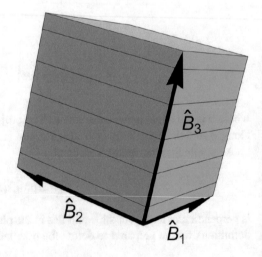

Fig. 12.7 Three noncolinear points (p_1, p_2, p_3) chosen on a single layer of anisotropy in the material

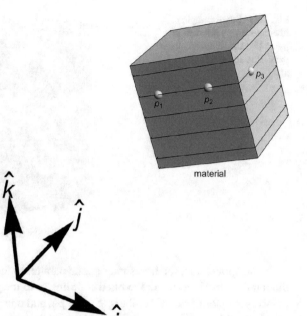

material

fixed coordinates

$$\vec{v} = \vec{q} - \text{proj}_u\left(\vec{q}\right). \tag{12.16}$$

This vector lies in the plane defined by the three points and is perpendicular to $\widehat{B}_1$. The unit vector for the y-axis of the jig will be

$$\widehat{B}_2 = \frac{\vec{v}}{\sqrt{\vec{v}\cdot\vec{v}}}. \tag{12.17}$$

Finally, define the z-axis of the jig to be

$$\widehat{B}_3 = \widehat{B}_1 \times \widehat{B}_2. \tag{12.18}$$

The vectors $\widehat{B}_1$, $\widehat{B}_2$, and $\widehat{B}_3$ are the column vectors of $\mathcal{R}$, where $\mathcal{R}$ is the rotation matrix needed to transform $\mathcal{F}_m$ into $\mathcal{F}_{\text{jig}}$.

Example 2: Fibrous Sample

If the material has fibers oriented parallel to one another, like the fibers in Dragon Skin shown in Fig. 12.2, place a small cube of this in the jig so that the fibers align with the x-axis (Fig. 12.8).

Fig. 12.8 Small sample of fiber anisotropy oriented in a jig so that the fibers are parallel to the x-axis (B_1 axis)

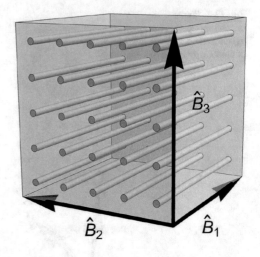

In the application, the fibers may not be parallel to the fixed coordinates. To find the rotation matrix to rotate the fixed coordinates to the jig coordinates, choose two points lying along one of the fibers, $\vec{p}_1$ and $\vec{p}_2$, and one point on another fiber in the x-y plane of the jig, $\vec{p}_3$ (Fig. 12.9). Then follow the same steps as in Example 1. This will result in the rotation matrix we need, $\mathcal{R}$.

Finding $\mathcal{T}$ from $\mathcal{F}$ (Gram-Schmidt QRD)

Any deformation of the material can be expressed locally in terms of the fixed coordinates as a deformation matrix, $\mathcal{F}_m$. This deformation matrix must be first transformed to the jig coordinates before ϵ can be calculated. The transformation is carried out with the $\mathcal{R}$ matrix described in the last two sections. The result is the deformation in terms of the jig coordinates, $\mathcal{F}_{\text{jig}}$. The process is a rotation of coordinates defined by (see Appendix A) the following:

$$\mathcal{F}_{\text{jig}} = \mathcal{R}^T \cdot \mathcal{F}_m \cdot \mathcal{R}. \tag{12.19}$$

Note that if the jig and fixed coordinates are aligned, $\mathcal{R}$ is the identity matrix, and

$$\mathcal{F}_{\text{jig}} = \mathcal{F}_m. \tag{12.20}$$

Now transform $\mathcal{F}_{\text{jig}}$ into the upper diagonal elements of the $\mathcal{T}$ matrix. The Gram-Schmidt QRD algorithm can be used to do this (see Appendix B). Alternatively, if the column vectors of $\mathcal{F}_{\text{jig}}$ are $\vec{a}$, $\vec{b}$, and $\vec{c}$, then

Fig. 12.9 Two points lying along one fiber, $\vec{p}_1$ and $\vec{p}_2$, and one other point, $\vec{p}_3$, not on that fiber

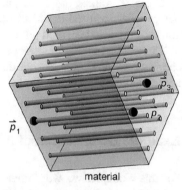

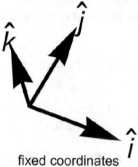

fixed coordinates

$$\mathcal{T}_{11} = \frac{\vec{a} \circ \vec{a}}{\sqrt{\vec{a} \circ \vec{a}}}$$

$$\mathcal{T}_{12} = \frac{\vec{a} \circ \vec{b}}{\sqrt{\vec{a} \circ \vec{a}}}$$

$$\mathcal{T}_{13} = \frac{\vec{a} \circ \vec{c}}{\sqrt{\vec{a} \circ \vec{a}}}$$

$$\mathcal{T}_{22} = \frac{\sqrt{\left(\vec{a} \times \vec{b}\right) \circ \left(\vec{a} \times \vec{b}\right)}}{\sqrt{\vec{a} \circ \vec{a}}}$$

$$\mathcal{T}_{23} = \frac{\left(\vec{a} \times \vec{b}\right) \circ \left(\vec{a} \times \vec{c}\right)}{\sqrt{\left(\vec{a} \circ \vec{a}\right)\left(\vec{a} \times \vec{b}\right) \circ \left(\vec{a} \times \vec{b}\right)}}$$

$$\mathcal{T}_{33} = \frac{\sqrt{\left(\left(\vec{a} \times \vec{b}\right) \circ \vec{c}\right)^2}}{\sqrt{\left(\vec{a} \times \vec{b}\right) \circ \left(\vec{a} \times \vec{b}\right)}}.$$

(12.21)

These $\mathcal{T}$ values are invariant under rotations after a deformation, but not before – as required of anisotropy. These six values, $\mathcal{T}_{11}$, $\mathcal{T}_{12}$, $\mathcal{T}_{13}$, $\mathcal{T}_{22}$, $\mathcal{T}_{23}$, and $\mathcal{T}_{33}$, are also the input values which define ϵ.

There are a number of steps in this process, so a simple example using a simple spring model of a homogeneous deformation of an anisotropic sample is given in the next section to help clarify the process.

Example Deformation of a Simple Spring Model

For this example, we will simulate the homogeneous deformation of a cube by assuming its anisotropy can be described by a simple spring model. To create the model, define the four nodes,

$$
\begin{aligned}
\vec{p}_1 &= (0, 0, 0) \\
\vec{p}_2 &= (1, 0, 0) \\
\vec{p}_3 &= (0, 1, 0) \\
\vec{p}_4 &= (0, 0, 1).
\end{aligned}
\tag{12.22}
$$

Connect these four nodes by linear elastic springs with spring constants, k (Fig. 12.10).

When the "material" is deformed, each node, $\vec{p}_i$, moves to a new location, $\vec{p}_i'$. The energy associated with this model is

$$
\epsilon = 1/2k \sum_{i=1}^{4} \sum_{j=1}^{i} \left(\sqrt{\left(\vec{p}_i' - \vec{p}_j' \right)^2} - \sqrt{\left(\vec{p}_i' - \vec{p}_j' \right)^2} \right)^2.
\tag{12.23}
$$

A deformation would transform the $\vec{p}_i$ vectors into $\vec{p}_i'$ vectors by a multiplication of $\mathcal{F}$ plus a constant vector $\vec{b}$, i.e.,

$$
\vec{p}_i' = \mathcal{F}\vec{p}_i + \vec{b}.
\tag{12.24}
$$

If the material is placed in a measurement jig, the $\vec{p}_i'$ vectors would be restricted to

$$
\begin{aligned}
\vec{p}_1' &= (0, 0, 0) \\
\vec{p}_2' &= (1 + dxx, 0, 0) = (\mathcal{T}_{11}, 0, 0) \\
\vec{p}_3' &= (dxy, 1 + dyy, 0) = (\mathcal{T}_{12}, \mathcal{T}_{22}, 0) \\
\vec{p}_4' &= (dxz, dyz, 1 + dzz) = (\mathcal{T}_{13}, \mathcal{T}_{23}, \mathcal{T}_{33})
\end{aligned}
\tag{12.25}
$$

and

Fig. 12.10 Springs connecting $\vec{p}_1$, $\vec{p}_2$, $\vec{p}_3$, and $\vec{p}_4$ to simulate a unit cube of material

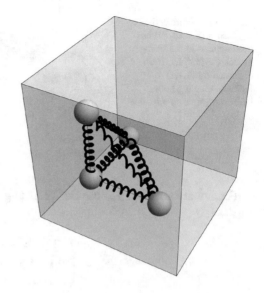

$$\vec{b} = 0 \tag{12.26}$$

so that

$$\mathcal{F} = \mathcal{F}_{\text{jig}} = \mathcal{T} = \begin{pmatrix} \mathcal{T}_{11} & \mathcal{T}_{12} & \mathcal{T}_{13} \\ 0 & \mathcal{T}_{22} & \mathcal{T}_{23} \\ 0 & 0 & \mathcal{T}_{33} \end{pmatrix} \tag{12.27}$$

so that $\mathcal{F}_{\text{jig}}$ is an upper triangular matrix.

Substituting Eq. 12.25 into Eq. 12.23 gives the energy function in terms of the elements of $\mathcal{T}$ as

$$\varepsilon = 1/2k\left((\mathcal{T}_{11} - 1)^2 + \left(\sqrt{\mathcal{T}_{12}^2 + \mathcal{T}_{22}^2} - 1\right)^2 + \left(\sqrt{(\mathcal{T}_{12} - \mathcal{T}_{11})^2 + \mathcal{T}_{22}^2} - \sqrt{2}\right)^2 \right.$$
$$+ \left(\sqrt{\mathcal{T}_{13}^2 + \mathcal{T}_{23}^2 + \mathcal{T}_{33}^2} - 1\right)^2 + \left(\sqrt{(\mathcal{T}_{13} - \mathcal{T}_{11})^2 + \mathcal{T}_{23}^2 + \mathcal{T}_{33}^2} - \sqrt{2}\right)^2$$
$$\left. + \left(\sqrt{(\mathcal{T}_{13} - \mathcal{T}_{12})^2 + (\mathcal{T}_{23} - \mathcal{T}_{22})^2 + \mathcal{T}_{33}^2} - \sqrt{2}\right)^2 \right).$$

$$\tag{12.28}$$

This is the energy function of our model in terms of the upper triangular matrix elements.

Now let us place our "material" into a simulation. First assume that the jig coordinates align with the fixed coordinates of the simulation. Since the jig and fixed coordinates are aligned, $\mathcal{F}_{\text{jig}}$ and $\mathcal{F}_m$ are the same, and Eq. 12.20 applies.

Fig. 12.11 Before and after $\mathcal{F}_m$ matrix (Eq. 12.29) has been applied to the spring model. ("After" has been displaced by (1,1,1) to facilitate viewing both before and after on the same plot)

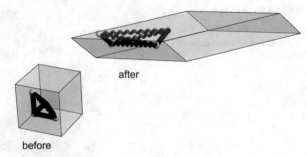

after

before

For any deformation, $\mathcal{F}_m$, use Eq. 12.21 to get the elements of $\mathcal{T}$. For example, suppose

$$\mathcal{F}_m = \begin{pmatrix} 2 & 2 & -0.4 \\ 0.1 & 1.5 & -0.3 \\ 0.2 & 0.3 & 0.5 \end{pmatrix}. \tag{12.29}$$

Then, Fig. 12.11 results.
Applying Eq. 12.21 to $\mathcal{F}_m$ gives

$$\mathcal{T} = \begin{pmatrix} 2.0 & 2.1 & -.36 \\ 0 & 1.4 & -.24 \\ 0 & 0 & 0.56 \end{pmatrix} \tag{12.30}$$

and the energy of deformation by Eq. 12.28 is

$$\text{energy} = 3.5 \, k. \tag{12.31}$$

As a second example, place the same "material" into a simulation so that the anisotropy is rotated 30 degrees about the z-axis. Apply the same deformation as in the last example, $\mathcal{F}_m$ (Eq. 12.29). The result is shown In Fig. 12.12.

To calculate the $\mathcal{T}$ values in terms of jig coordinates, we must first transform $\mathcal{F}_m$ which is in fixed coordinates into jig coordinates by rotating $\mathcal{F}_m$ about the z-axis by 30 degrees:

$$\mathcal{R} = \begin{pmatrix} \cos\left(30^\circ\right) & -\sin\left(30^\circ\right) & 0 \\ \sin\left(30^\circ\right) & \cos\left(30^\circ\right) & 0 \\ 0 & 0 & 1 \end{pmatrix}, \tag{12.32}$$

then

Fig. 12.12 Deformation of cube by $\mathcal{F}_m$, Eq. 12.28, but now the spring model is initially rotated by $20°$ about the z-axis. ("After" has been displaced by (1,1,1) to facilitate viewing both before and after on the same plot)

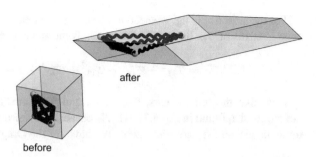

after

before

$$\mathcal{F}_{\text{jig}} = \mathcal{R}^T \mathcal{F} \mathcal{R} \tag{12.33}$$

and Eq. 12.21 applied to $\mathcal{F}_{\text{jig}}$ gives

$$\mathcal{T} = \begin{pmatrix} 2.8 & 1.1 & -0.41 \\ 0 & 0.98 & -0.15 \\ 0 & 0 & 0.56 \end{pmatrix}. \tag{12.34}$$

The energy of deformation is then

$$\text{energy} = 4.1 \text{ k}. \tag{12.35}$$

Note that this energy is different than Eq. 12.31 even though we used the same energy function, ϵ, and the same deformation, $\mathcal{F}_m$. This difference occurs because the anisotropy was initially rotated in the second example, but not in the first.

Numerical Simulation Example

In this example, anisotropy is added to a cylinder made of Dragon Skin material by adding elastic fibers. We will assume that the fibers are very thin, contributing little to the total volume, but stiffer than the Dragon Skin material. The resulting aniso-tropic ϵ will be the sum of the Dragon Skin energy and the energy of the elastic fibers. If we place this material in a jig, we can construct ϵ in terms of $\mathcal{T}_{ij}$ as we did with the last example.

The Dragon Skin energy is (Eq. 10.13)

$$\begin{aligned} \varepsilon_{\text{dragon–skin}} &= b(\mathcal{I}_1 - 3) \\ \text{where} \quad b &= 3.4 \times 10^4 \text{ N/m}^2 \end{aligned} \tag{12.36}$$

with the Dragon Skin material assumed to be incompressible.

For the anisotropic jig, we need to construct the energy function of a Dragon Skin cube in terms of $\mathcal{T}_{ij}$. In jig coordinates $\mathcal{T}_{ij}$ is the same as $\mathcal{F}_{ij}$. Thus, for I_1,

$$\mathcal{I}_1 = \mathcal{T}_{11}^2 + \mathcal{T}_{12}^2 + \mathcal{T}_{13}^2 + \mathcal{T}_{22}^2 + \mathcal{T}_{23}^2 + \mathcal{T}_{33}^2. \qquad (12.37)$$

Note that the three shears, $\mathcal{T}_{21}$, $\mathcal{T}_{31}$, and $\mathcal{T}_{32}$, are missing from the general definition of I_i found in Eq. 5.21. This is because the anisotropic jig does not change these values, so they are set to zero. We have for the Dragon Skin energy,

$$\epsilon_{\text{dragon-skin}} = b\big(\mathcal{T}_{11}^2 + \mathcal{T}_{12}^2 + \mathcal{T}_{13}^2 + \mathcal{T}_{22}^2 + \mathcal{T}_{23}^2 + \mathcal{T}_{33}^2 - 3\big). \qquad (12.38)$$

For the first simulation, we choose the fibers to be aligned with the x-axis in the jig. This gives from Eq. 4.23

$$\epsilon_{\text{fibers}} = c(\mathcal{T}_{11} - 1)^2. \qquad (12.39)$$

Combining these two energies gives

$$\epsilon_{\text{fibers}} + \epsilon_{\text{dragon-skin}} = c(\mathcal{T}_{11} - 1)^2 \\ + b\big(\mathcal{T}_{11}^2 + \mathcal{T}_{12}^2 + \mathcal{T}_{13}^2 + \mathcal{T}_{22}^2 + \mathcal{T}_{23}^2 + \mathcal{T}_{33}^2 - 3\big). \qquad (12.40)$$

Lastly, we need to enforce the condition of incompressibility. This is done in this simulation in same manner as in Chaps. 7 and 8. That is, an energy term, $a\,(I_3 - 1)^2$, with $a \gg b$ and $a \gg c$, is added to Eq. 12.40 to enforce near incompressibility:

$$\varepsilon = c(\mathcal{T}_{11} - 1)^2 + b\big(\mathcal{T}_{11}^2 + \mathcal{T}_{12}^2 + \mathcal{T}_{13}^2 + \mathcal{T}_{22}^2 + \mathcal{T}_{23}^2 + \mathcal{T}_{33}^2 - 3\big) \\ + a\big(\mathcal{I}_3 - 1\big)^2. \qquad (12.41)$$

In jig coordinates $\frac{\partial x_2}{\partial X_1}$, $\frac{\partial x_3}{\partial X_1}$, $\frac{\partial x_3}{\partial X_2}$ are zero so that the general I_3 found in Eq. 5.21 reduces to

$$\mathcal{I}_3 = \left(\frac{\partial x_1}{\partial X_1}\right)\left(\frac{\partial x_2}{\partial X_2}\right)\left(\frac{\partial x_3}{\partial X_3}\right) = \mathcal{T}_{11}\mathcal{T}_{22}\mathcal{T}_{33}. \qquad (12.42)$$

In the following simulation, computer units are chosen so that b is 1, c is chosen to be 10, and a is chosen to be 100. These are chosen so that the value of "a" will enforce near incompressibility and the value of c will be larger than the value of b so that the fibers are "stiffer" than the Dragon Skin material. In this case, the energy of the Dragon Skin material with fibers parallel to the x-axis is given by Eq. 12.43.

Fig. 12.13 The original orientation of the material in the darker blue and the final condition of the material is shown in the light blue. The original cylinder has been compressed by 20% to produce the final condition of the material. The final condition is elliptical instead of circular

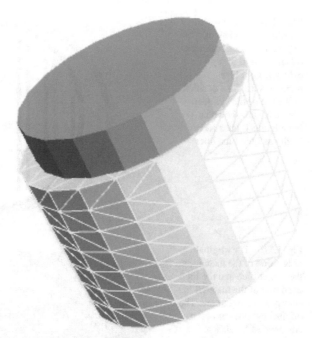

$$\epsilon = 10(\mathcal{T}_{11} - 1)^2 + 1\left(\mathcal{T}_{11}{}^2 + \mathcal{T}_{12}{}^2 + \mathcal{T}_{13}{}^2 + \mathcal{T}_{22}{}^2 + \mathcal{T}_{23}{}^2 + \mathcal{T}_{33}{}^2 - 3\right)$$
$$+ 100(\mathcal{T}_{11}\mathcal{T}_{22}\mathcal{T}_{33} - 1)^2 \tag{12.43}$$

In the first simulation, the fibers in the cylinder are oriented perpendicular to the axis of the cylinder. The cylinder is compressed by moving the top nodes downward in the same manner as the deformation in section "Cylindrical compression test" in Chap. 8. The sides are allowed to expand similar to simulation section "Cylindrical compression test" in Chap. 8. The z component of the top nodes was decreased in five steps until the cylinder is 80% of its original height. The bottom nodes were fixed in the z direction. Both the top and bottom nodes were allowed to move freely in the x and y directions, except for the center node of the top and bottom which were fixed in x and y. The result of this simulation is shown in Fig. 12.13.

The material is not extended as much along the fiber direction as perpendicular to the fibers. This produces the oval appearance of the compressed cylinder. A top view of this is cartooned in Fig. 12.14. The material bulged out perpendicular to the fiber alignment so as to minimize the change in the length of the fibers, yet maintain a constant volume as the cylinder is compressed.

As a second example, keep everything the same, except rotate the fibers $70°$ about the y-axis. The result is shown in Fig. 12.15.

To understand the final condition of the cylinder in this second example, it is helpful to draw a cartoon of a cross section of the cylinder before and after compression (Fig. 12.16).

Fig. 12.14 Cartoon
showing a top view of
Fig. 12.13. Black indicates
initial position of top of
cylinder; light blue
represents the final position
of the top of the cylinder.
The fibers are shown as
initially black lines and
finally as light blue lines

Fig. 12.15 The original
orientation of the material in
the darker blue and the final
condition of the material is
shown in the light blue. The
original cylinder has been
compressed by 20% to
produce the final condition
of the material. Note that the
cylinder is deformed by
rotation as well as
compression

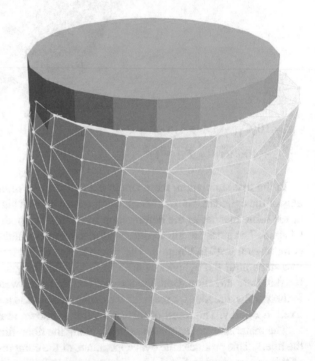

The material bulged out in Fig. 12.15 to allow the fibers to rotate to try to maintain their length. Even though the material has the same ϵ, there is a great difference in the response shown in Figs. 12.13 and 12.15, because the initial anisotropy is oriented differently in the two cases. This illustrates how essential it is that the initial orientation of the anisotropy be included in the simulation as well as the anisotropic energy function, ϵ.

Fig. 12.16 Cartoon
showing a cross sectional
view of Fig. 12.15. Black
indicates initial position of
the outline of the cylinder
and the fibers causing the
anisotropy. Light blue
indicates the final
orientation of the cylinder
and the fibers. Note how the
fibers have rotated to
maintain their length as
much as possible. This has
produced the bulge at the
upper top and lower bottom
of the cylinder as shown in
Fig. 12.15

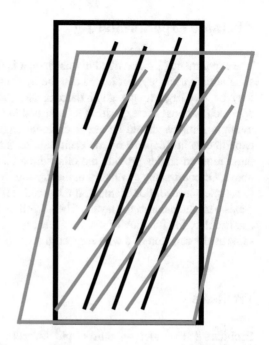

Alternate Set of Anisotropic Invariants

The invariants for numerical calculations of anisotropy (Eq. 12.21) could be simplified to (see Problem 11)

$$\begin{aligned}
t_{11} &= \vec{a} \circ \vec{a} \\
t_{12} &= \vec{a} \circ \vec{b} \\
t_{13} &= \vec{a} \circ \vec{c} \\
t_{22} &= \left(\vec{a} \times \vec{b}\right) \circ \left(\vec{a} \times \vec{b}\right) \cdot \\
t_{23} &= \left(\vec{a} \times \vec{b}\right) \circ \left(\vec{a} \times \vec{c}\right) \\
t_{33} &= \left(\vec{a} \times \vec{b}\right) \circ \vec{c}
\end{aligned} \tag{12.44}$$

This would speed up numerical calculations but would complicate the expression of ϵ in terms of experimental measurements. In programs, I have opted to stay with Eq. 12.21.

Alternate Experimental Jig

The experimental jig described in this chapter is described to clarify the approach to simulating anisotropy, but in practice the jig frictional forces on the sides of the jig may be too large to get good deformation force measurements. An alternative approach is to use the experiments described in Chap. 10. To map the entire space, multiple samples should be cut at different angles to the anisotropy and then the experiments in Chap. 10 run on each sample. Since there are three rotation angles, these rotation angles provide the other three independent parameters necessary to map ϵ. Of course, any symmetry in the anisotropy can be used to reduce the number of samples needed. For example, if a layered anisotropy is as in Fig. 12.6, orient the z-axis perpendicular to the layers. Then ϵ will be independent of rotations about the z-axis. Also, the function of ϵ with respect to the rotation angle about x will be the same as the function of ϵ with respect to the rotation angle about y.

Problems

Problem 1 Consider an anisotropic layer to be defined by the three points, $(0.93,1.1,1.7)$, $(1.4,1.2,1.9)$, and $(1.6,1.5,1.9)$. Find the rotation matrix $\mathcal{R}$ needed to transform from the jig coordinates to the fixed coordinates.

Problem 2 Assume we have a material made up of just parallel fibers (parallel linear elastic springs), so that there are five fibers per m^2. Each fiber has a spring constant of 2 N/m. Assume the material that the fibers are embedded in has a spring constant of 0. If we place this material in a jig with the fibers aligned with the x-axis, what will be the energy function, ϵ, in terms of the elements of $\mathcal{T}$?

Problem 3 Assume we have a material made up of just parallel fibers (parallel linear elastic springs), so that there are five fibers per m^2. Each fiber has a spring constant of 2 N/m. Assume the material that the fibers are embedded in has a spring constant of 0. If we place this material in a jig with the fibers aligned with the z-axis, what will be the energy function, ϵ, in terms of the elements of $\mathcal{T}$?

Problem 4 Show that if we start with the fibers of Problem 2 aligned in the x direction and then rotate the material so that the fibers are now in the z direction, we get the same energy of deformation that we did in Problem 3.

Problem 5 Repeat the calculations with the spring model in Fig. 12.10 and verify Eq. 12.28.

Problem 6 Start with Eq. 12.29. Then (a) verify 12.30 using Eq. 12.21, and (b) verify 12.31 using Eq. 12.28.

Problem 7 Given the deformation $\mathcal{F} = \begin{pmatrix} 2 & 2 & -0.4 \\ 0.1 & 1.5 & -0.3 \\ 0.2 & 0.3 & 0.5 \end{pmatrix}$, show that a rotation

of $\mathcal{R} = (\cos\theta \ \ 0 \ \ 0 \ 1 \ 0 \ \sin 0 \ \cos\ )$ with $\theta = 25°$ after the $\mathcal{F}$ deformation does not change the $\mathcal{T}_{ij}$ values from Eq. 12.30.

Problem 8 Assume the planes of anisotropy in Fig. 12.7 are defined by $z = 2x + 3y + c$, where $c = 1$ through 5. Assume the measurement jig has these planes lying in the x-y plane of the jig. Find the rotation matrix that will rotate the fixed coordinates into the jig coordinates.

Problem 9 Find a vector perpendicular to $\vec{v}$ lying in the plane defined by $\vec{v}$ and $\vec{s}$, where

$$\vec{v} = (1, 3, -2)$$

and

$$\vec{s} = (2, 1, 1).$$

Problem 10 Show that $\mathcal{I}_3 = \left(\frac{\partial x_1}{\partial X_1}\right)\left(\frac{\partial x_2}{\partial X_2}\right)\left(\frac{\partial x_3}{\partial X_3}\right)$ if $\frac{\partial x_2}{\partial X_1}$, $\frac{\partial x_3}{\partial X_1}$, $\frac{\partial x_3}{\partial X_2}$ are all zero using the definition of $\mathcal{I}_3$ in Eq. 5.21.

Problem 11 Show that the alternate invariants expressed in Eq. 12.44 follow directly from the invariant property of the elements of $\mathcal{T}$ as described in Eq. 12.21. (Hint: Solve the first equation in Eq. 12.21 for for $\vec{a}.\vec{a}$ in terms of $\mathcal{T}_{11}$. Since $\mathcal{T}_{11}$ is invariant and $\vec{a} \circ \vec{a} = f(\mathcal{T}_{11})$, $\vec{a} \circ \vec{a}$ is also invariant. Continue with other alternate invariant terms.)

Problem 12 Show that the deformation $\mathcal{F}$ given in Eq. 12.29 produces the same energy of deformation for the two results Eqs. 12.31 and 12.35 by using the points equation, Eq. 12.22 substituted into Eq. 12.23.

Chapter 13
Plot Deformation, Displacements, and Forces

In this chapter, we will describe several post-processing functions. A file is written from the main simulation which saves the output of each time step. The processes described in this chapter read the output and display grid deformations, internal forces, boundary forces, and the displacement of the nodes. The equations for each of these, for both 2D and 3D, are included in this chapter. However, only 2D figure examples are shown here because it is difficult to interpret 3D post-processing figures without either rotation or time laps pictures – both of which are impossible on the printed page.

Deformation of Local Regions

We must first discuss exactly what we mean by a deformation. For example, we do not wish to consider a translation of a deformation, even though the position of the body has been changed. Similarly, if we rotate the body after the deformation, the final and initial body can still be superimposed. This too is not a deformation.

The deformation information for each triangle or tetrahedron is contained in

$$\mathcal{F}_{ij} = \frac{\partial x_i}{\partial x_{j'}} \tag{13.1}$$

but in the absence of a deformation,

$$\mathcal{F}_{ij} = \delta_{ij} \tag{13.2}$$

Supplementary Information The online version contains supplementary material available at [https://doi.org/10.1007/978-3-031-09157-5_13].

so that $\mathcal{F}_{ij}$ is not zero in that case. In addition, the rotation of the body after the deformation results in a change in $\mathcal{F}_{ij}$. Therefore, we must extract the deformation from $\mathcal{F}_{ij}$. To do this, first take the Gram-Schmidt QRD of $\mathcal{F}_{ij}$ found in Appendix A. This gives

$$\mathcal{F}_{ij} = \mathcal{R}_{ik}\mathcal{T}_{kj}, \tag{13.3}$$

where $\mathcal{T}_{kj}$ is an upper triangular matrix and $\mathcal{R}_{ik}$ is a rotation matrix.

Define the deformation matrix, $\mathcal{D}_{ij}$, as

$$\mathcal{D}_{ij} = \mathcal{R}_{ik}\left(\mathcal{T}_{kj} - \delta_{kj}\right). \tag{13.4}$$

To display $\mathcal{D}_{ij}$, define the components of the vectors $\rightarrow dv_x$ and $\rightarrow dv_x$ as follows:

$$\begin{aligned}(dv_x)_i &= \mathcal{D}_{i1}\\(dv_y)_i &= \mathcal{D}_{i2}\\&\text{for } i = 1 \text{ or } 2,\end{aligned} \tag{13.5}$$

where

$(dv_x)_i$ is the ith component of a vector showing the displacement of the x face of a unit cube within the region

$(dv_y)_i$ is the ith component of a vector showing the deformation of the y face of a unit cube within the region.

In 2D, the deformation is plotted as two double-headed arrows. These arrows are twice the length of the vectors, $(dv_x)_i$ and $(dv_x)_i$. The center of each of these is the center of the region for which the deformation has been computed. If the deformation is extensional, the arrowheads on the end of the double-headed arrows point outward. If the deformation is compressional, the arrowheads point inward. If the deformation is pure shear with no extension or compression, the arrow has no point on the end. The arrows can be scaled by the user so that the arrows lie within the simulated grids. These arrows are colored as red for deformations in the x (i.e., $j = 1$) and blue for deformations in the y (i.e., $j = 2$).

Fig. 13.1 A single grid extended in the x direction. The double-headed arrow expresses this deformation

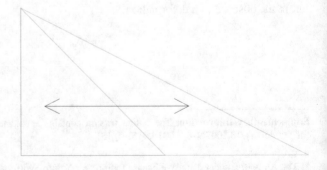

Fig. 13.2 An extension
only in the x direction as in
Fig. 13.1 followed by a
rotation of the grid

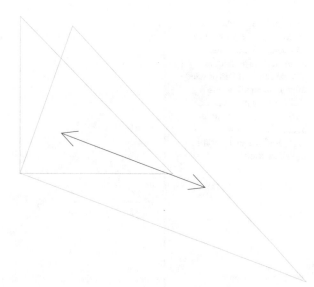

Figure 13.1 shows the deformation arrows for a single grid which has been extended in the x direction (horizontal), but unchanged in the y direction (vertical). The initial triangle is gray, and the triangle after deformation is red. Figure 13.2 shows the rotation of the final nodes after the same deformation in Fig. 13.1. No change in the deformation has been made relative to the deformed body. The deformation arrow just rotates with the deformed body. Figure 13.3 shows the deformation arrows for a single grid which has been compressed in the x direction and extended in the y direction. Figure 13.4 shows a single grid which has been sheared, with the deformation defined as

$$x = X + 0.3Y$$
$$y = Y \tag{13.6}$$

As a last single grid example, Fig. 13.5 shows a single triangle deformed using

$$x = 0.8X + 0.1Y$$
$$y = 0.4X + 1.3Y. \tag{13.7}$$

Figure 13.6 shows the 2D compression of a rectangle, holding the top and bottom fixed in the x direction. (This is a 2D representation of the deformation of Fig. 8.3.) Note that the forces on the y surface (blue) are mostly compressional and that the forces on the x face (red) are mostly extensional.

In 3D the vectors $(dv_x)_i$, $(dv_y)_i$, and $(dv_z)_i$ are the deformation of the x, y, and z faces, respectively, of a unit cube within each tetrahedron. These are defined as

Fig. 13.3 A single grid extended in the y direction and compressed in the x direction. The double-headed red arrow expresses the compression in the x direction with reversed arrowheads. The double-headed blue arrow expressed the extension in the y direction

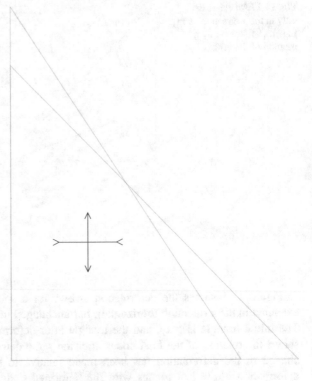

Fig. 13.4 A single grid extended in shear with the top node moved parallel to the x-axis. The blue line indicates the y face only in the x direction with no compression or extension in the y direction

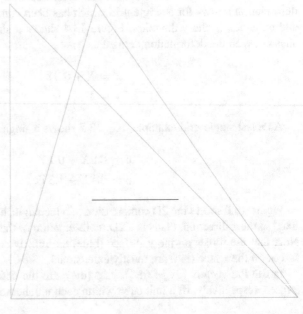

Fig. 13.5 A single grid extended and sheared in the y direction, compressed and sheared in the x direction. The blue arrow indicates the extension of the y face, and the red arrow indicates the compression of the x face

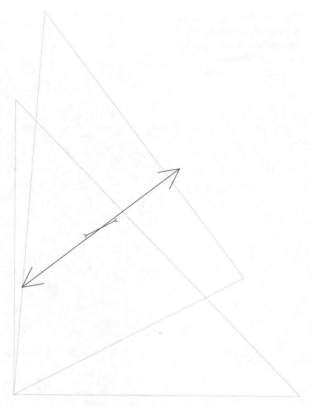

$$(dv_x)_i = \mathcal{D}_{i1}$$
$$(dv_y)_i = \mathcal{D}_{i2}$$
$$(dv_z)_i = \mathcal{D}_{i3} \tag{13.8}$$
$$\text{for } i = 1, 2, \text{ and } 3,$$

where

$(dv_x)_i$ is the ith component of a vector showing the displacement of the x face of a unit cube within the region
$(dv_y)_i$ is the ith component of a vector showing the deformation of the y face of a unit cube within the region
$(dv_z)_i$ is the ith component of a vector showing the deformation of the z face of a unit cube within the region.

In 3D the deformation of each grid can be plotted as three double-headed arrows. These arrows are twice the length of the vectors, $(dv_x)_i$, $(dv_y)_i$, and $(dv_y)_i$. The center of each of these arrows is the center of each tetrahedron. If the deformation is extensional, the arrowhead on the end of the double-headed arrow point outward. If the deformation in compressional, the arrows point inward. If the deformation is

Fig. 13.6 The compression
of a rectangle showing the
deformation of each grid in
the x and y directions

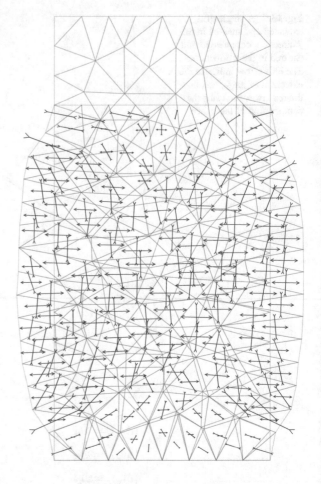

pure shear with no extension or compression, the arrow has no point on the end. The
arrows can be scaled by the user so that the largest arrow lies within its tetrahedron.
These arrows are colored as red for $j = 1$, green for $j = 2$, and blue for $j = 3$.

Internal Forces on Local Regions

The forces per unit initial area in 3D (or initial length in 2D) are defined as

$$\mathcal{P}_{ij} = \frac{\partial \epsilon}{\partial \left(\partial x_i / \partial x_j\right)} \tag{13.9}$$

The values of $\mathcal{P}_{ij}$ are displayed in the same manner as $\mathcal{D}_{ij}$ values shown in
Eq. 13.5 except magenta is used for $(dv_x)_i$ and cyan is used for $(dv_y)_i$.

Fig. 13.7 A single grid
extended in the x direction
as was done in Fig. 13.1.
The triangle is nearly
incompressible, so that force
is required in both the x and
y directions. The magenta
force per unit initial area
displays the force in the
x direction. The cyan
double-headed arrow
displays the force in the
y direction

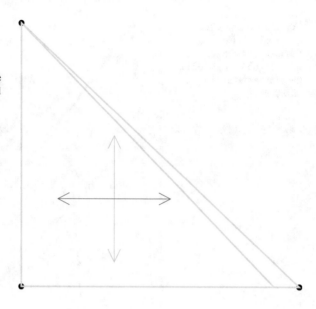

Fig. 13.8 A single grid
extended in the x direction
as in Fig. 13.1. The triangle
changes shape to maintain a
nearly constant volume, so
that force is required only in
the x direction. The magenta
double-headed arrow
displays the force per unit
initial length in the
x direction

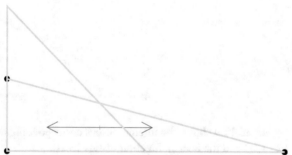

Figure 13.7 shows the same extension as Fig. 13.1 with the energy function

$$\epsilon = \mathcal{I}_1 + 100(\mathcal{I}_3 - 3)^2. \tag{13.10}$$

This energy function tries to maintain a constant volume. As a result, an extensional force is required to keep the y direction from compressing as the x direction increases.

If we allow the y direction to decrease to maintain a constant volume, the extensional force in the y direction is decreased. Figure 13.8 shows this result. (The final volume is not quite exactly preserved because the multiple of the $(I_3 - 1)^2$ term would have to approach an infinite value for this to be exactly the case.)

Fig. 13.9 A single grid
deformed as in Fig. 13.5, but
this time the force per unit
initial length is displayed in
the x (magenta) and y (cyan)
directions

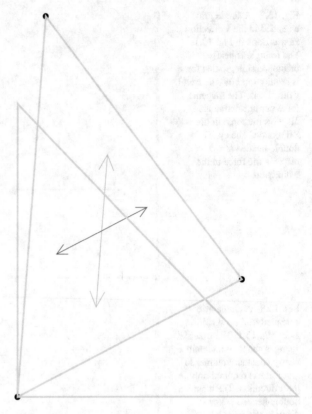

Figure 13.9 shows the internal forces corresponding to the deformation shown in
Eq. 13.7 and the energy function of Eq. 13.10.

Figure 13.10 shows the internal force per unit initial length for the same deformation as shown in Fig. 13.6. Note that there are large compressive forces in the
y direction (cyan) for all the triangles. Large compressive forces are shown in the
x direction (magenta) only for the grids at the top and bottom, where the grids are
constrained not to move in the x (horizontal) direction.

Displacement of Nodes

Displacements for each node are defined as

$$u_i = x_i - X_i. \tag{13.11}$$

Fig. 13.10 The
compression of a rectangle
showing the force per unit
length of each grid in the
x and y directions

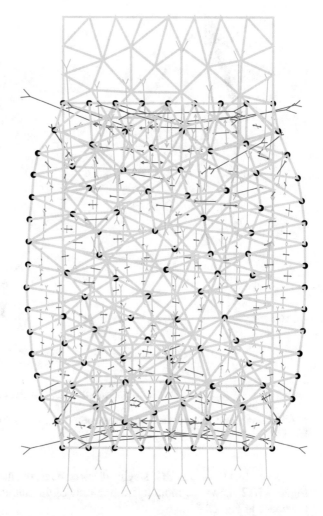

Equation 13.11 can be used to plot the displacement of each node from the initial
to the final location as a single arrow, but a more useful plot is to define a series of
line segments ending in an arrowhead. The first line segment from x_i to u_i is defined
where X_i is the initial point location and x_i is the location of the node after the first
step. The next line segment from x_i to u_i is defined with X_i the location of the node
after the first step and x_i the location of the node after the second step. This is
repeated over and over until the last step is reached. Then an arrowhead is placed on
the last line in the direction of motion.

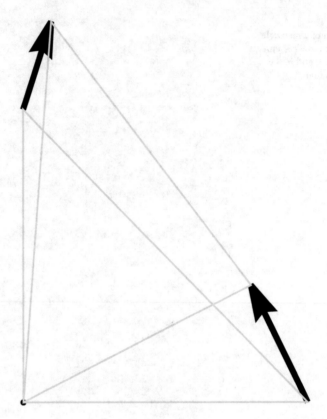

Fig. 13.11 A single grid showing the displacement of the nodes for the deformation shown in Fig. 13.5

Figure 13.11 shows the single displacement of the nodes in Fig. 13.5. Figure 13.12 shows the more interesting case of the multiple step displacement of the nodes in Fig. 13.6.

Boundary Forces

The most straight forward method of computing boundary forces is to compute the force at each node using

$$F_i = \frac{\partial}{\partial x_i} \sum_{k=1}^{N_p} \epsilon_k (V_0)_k, \qquad (13.12)$$

where

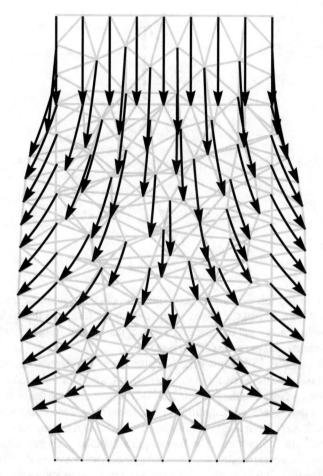

Fig. 13.12 The displacement of nodes is shown for the compression of the rectangle shown in Fig. 13.6

ϵ_k = energy per unit volume of triangle (tetrahedron) k
$(V_0)_k$ = initial area (volume) of triangle (tetrahedron) k.

The sum is over all the triangles (tetrahedra) which contain the node of interest. As a result of minimizing the energy, the force will be zero for all nodes except for fixed nodes.

Examples which display the forces are found in Figs. 13.13, 13.14, 13.15, and 13.16.

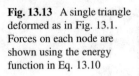

Fig. 13.13 A single triangle
deformed as in Fig. 13.1.
Forces on each node are
shown using the energy
function in Eq. 13.10

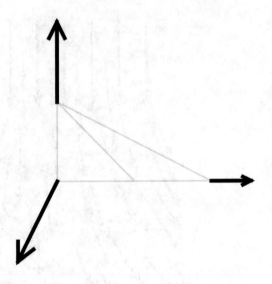

The sum of the forces must be zero since the object is stationary, as shown in Fig. 13.13. Also note that the forces computed are the forces *on* the nodes. If force per unit length or force per unit area are needed, these nodal forces must be divided by the associated length or area of the node.

For example, Fig. 13.14 shows the compression of a rectangle, allowing the top and bottom nodes to move freely in the *x* direction (except for one center node at the top and bottom). Note that the forces computed on each node are the same except for the corner nodes, where the magnitude of the forces are 1/2 the other nodes. This is because the force per unit length is the same everywhere on the top and bottom lines to give a uniform deformation. The line length associated with the corner nodes is 1/2 of the line length associated with the center nodes, so that when the force per unit length is computed, the force per unit length is the same all along the top and bottom of the rectangle – as expected.

Figure 13.15 shows the nodal forces for the deformation shown in Fig. 13.6. These nodal forces do not look correct. We would expect a uniform force distribution along the top and bottom nodes. The reason for this discrepancy is that the triangular regions used in the simulation are too large. If we increase the number of nodes so that the size of the triangular regions decreases, the node forces begin to make more sense.

Figure 13.16 shows the same deformation as in Fig. 13.15, but with more and more nodes. There are several changes that are occurring as the number of nodes increase. As we increase the number of nodes, we approach a solution which only

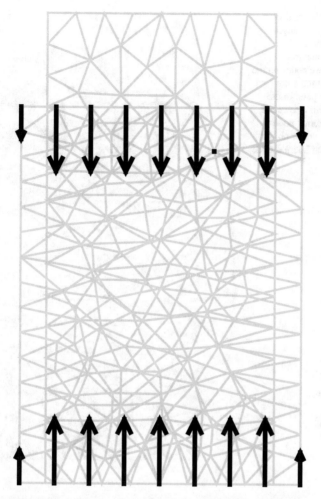

Fig. 13.14 The compression of Fig. 13.6, but this time allowing the x direction of the nodes to move freely so that the rectangle expands as it is compressed. The nodal forces at the top and bottom are shown. The nodal forces on the edges are 1/2 the other forces

has forces on the four corners of the rectangle. Also, the total force necessary to produce the deformation decreases as the number of grids increases. This decrease will continue as the number of grids increase to approach the bounding solution for an infinite number of grids. The total force on the bottom of the rectangle is plotted

Fig. 13.15 Forces shown on all nodes resulting from the compression of the rectangle shown in Fig. 13.6. Here nodes on the top and bottom are not allowed to expand in the x direction. No arrows are shown on internal nodes because the sum of all the forces on each internal node is zero

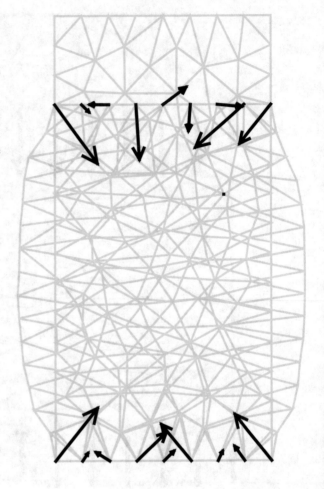

versus the number of grids used in the simulation in Fig. 13.17. Note that the total force on the bottom of the grid is approaching a fixed value as the number of nodes increase. Remember that we are simulating a continuous equation, Eq. 4.45, with a discontinuous grid. The equation is only fully satisfied as the number of grids become infinite.

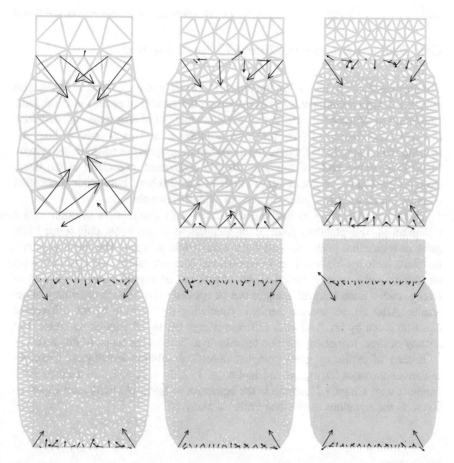

Fig. 13.16 Forces shown on all nodes resulting from the compression of the rectangle shown in Fig. 13.15 as the number of nodes in the simulation increases. The top left uses 35 nodes; the top middle, 104 nodes; the top right, 232 nodes; lower left, 449 nodes; lower center, 853 nodes; and lower right, 1394 nodes

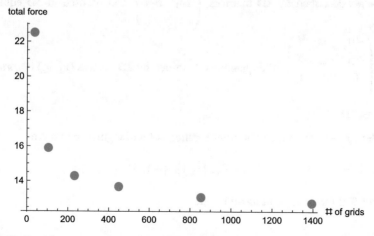

Fig. 13.17 Total force versus the number of nodes showing the approach to a final, "correct" force at the bottom of the rectangle with the deformation in Fig. 13.6

Grid Refinement and the Finite Element Method

The changes noted in increasing the number of grids lead us to a discussion of grid refinement. Grid refinement refers to increasing the number of nodes (and grids) in a simulation to increase the accuracy of the simulation. Grid refinement is often done in small areas instead of over the entire solution area. This is done to limit the increase in computational time that occurs when more grids are added to the system. There is an entire body of literature on grid refinement, especially in the linear elasticity literature. It is not our desire to delve into this here, but if your application requires higher accuracy, you will need to explore this technique.

There are other ways to increase the accuracy of a numerical solution with a minimum of computational time. In the simulations in this book, only linear triangular or tetrahedral grids are used for the numerical solution. The linear elasticity literature includes higher-order interpolations between nodes and different grids altogether – like rectangular grids. All of this can be applied here. In fact, all of what is called finite element analysis can be applied to increase the accuracy of the results. After all, all we are doing is simulating the solution of the differential equation given by Eq. 4.45 with different energy per initial volume functions. But a study of these numerical solution techniques is beyond the scope of this text.

Instead of exploring more complex numerical solution techniques, Chap. 14 re-derives the equations presented in Chaps. 1, 2, 3, 4, 5, and 6 in a much more compact way. Chapter 15 connects the equations to linear elasticity, and Chap. 16 connects the equations to classical finite elasticity.

Problems

The following problems are described only in 2D to simplify the mathematics. Quantities described by 2D matrices $\begin{pmatrix} a_{11} & a_{12} \\ a_{21} & a_{22} \end{pmatrix}$ can be used in 3D equations

with $\begin{pmatrix} a_{11} & a_{12} & 0 \\ a_{21} & a_{22} & 0 \\ 0 & 0 & 1 \end{pmatrix}$, and quantities described by 2D vectors (v_1, v_2) become $(v_1,$

$v_2, 0)$ in 3D.

Problem 1 Assume the initial point locations of a triangular region are

$$\{1, 2\}, \{3, 4\}, \{-1, 5\}$$

Find $\mathcal{F}$ if the final point locations are

$$\{1.7, 1\}, \{4.5, 1.8\}, \{0.4, 3.2\}$$

respectively.

Problem 2 Given

$$\mathcal{F} = \begin{pmatrix} .8 & -.5 \\ 0.3 & 1.2 \end{pmatrix}.$$

Find $\mathcal{R}$ and $\mathcal{T}$ from the Gram-Schmidt QRD process. The details of the process are found in Appendix B.

Problem 3 Given

$$\mathcal{R} = \begin{pmatrix} \cos\left[\frac{\pi}{9}\right] & \sin\left[\frac{\pi}{9}\right] \\ -\sin\left[\frac{\pi}{9}\right] & \cos\left[\frac{\pi}{9}\right] \end{pmatrix}$$

and

$$\mathcal{T} = \begin{pmatrix} 1.2 & 0.4 \\ 0 & 0.4 \end{pmatrix}.$$

Find the deformation matrix, $\mathcal{D}$.

Problem 4 Given the deformation matrix,

$$\mathcal{D} = \begin{pmatrix} 0.2 & 0.1 \\ -0.05 & -0.7 \end{pmatrix}.$$

Find the deformation vectors to be plotted.

Problem 5 Given the energy function

$$\epsilon = \mathcal{I}_1$$

and the deformation of

$$x_1 = 2X_1$$
$$x_2 = X_2,$$

Find $\mathcal{P}_{ij}$.

Problem 6 Given the following position of three points defining a triangular region, plot the initial and final triangle and the displacement arrows for all three points:

initial	step1	step2	step3
(1, 2)	(1.15, 2.1)	(1.2, 2.2)	(1.1, 2.5)
(1.5, 2.3)	(1.55, 2.3)	(1.6, 2.2)	(1.7, 2.0)
(1.2, 3.1)	(1.11, 3.1)	(1.3, 3.2)	(1.3, 3.0)

Problem 7 Given

$$\epsilon = \mathcal{I}_1.$$

Find the force on each point of the triangle initially defined by the three points, (1,1), (1.1,2), (2,2), and finally defined by the three points (in the same order), (1,1), (1.5,2.2), (2,2).

Problem 8 Consider a material with an anisotropic jig energy function, $\epsilon = (\mathcal{T}_{11} - 1)^2$. Orient the anisotropy so that it makes a 20-degree, clockwise angle from the y-axis. Define the nodes to be at the corners of the material, $X_1 = (0, 0)$, $X_2 = (1, 0)$, $X_3 = (0, 1)$, $X_4 = (1, 1)$. Define two triangles to cover the material: $\{X_1, X_2, X_3\}$ and $\{X_2, X_3, X_4\}$. Let the material be deformed so that $X_1 = (0, 0)$, $X_2 = (1, 0)$, $X_3 = (0, 0.8)$, $X_4 = (1, 0.8)$. Find the force per original length $\mathcal{P}_{ij} = \partial\epsilon/(\partial x_i/\partial X_j)$ for each of the two triangles making up the material. (ANS:
$$\mathcal{P}_{ij} = \begin{pmatrix} -0.0247 & -0.0677 \\ -0.0542 & -0.149 \end{pmatrix}).$$

Problem 9 Consider a material with an anisotropic jig energy function, $\epsilon = (\mathcal{T}_{11} - 1)^2$. Orient the anisotropy so that it makes a 20-degree, clockwise angle from the y-axis. Define the nodes to be at the corners of the material, $X_1 = (0, 0)$, $X_2 = (1, 0)$, $X_3 = (0, 1)$, $X_4 = (1, 1)$. Define two triangles to cover the material: $\{X_1, X_2, X_3\}$ and $\{X_2, X_3, X_4\}$. Let the material be deformed so that $X_1 = (0, 0)$, $X_2 = (1, 0)$, $X_3 = (0, 0.8)$, $X_4 = (1, 0.8)$. Find the force per original area $f_i = \partial\epsilon/\partial x_i$ for nodes X_3 and X_4. (ANS: $f_3 = (-0.0215, -0.0474)$ and $f_4 = (-0.0462, -0.102)$).

Problem 10 Use the results of Problems 8 and 9 to show that the same force is applied to the top of the square in both Problems 8 and 9.

Chapter 14
Euler-Lagrange Elasticity

The basic equations describing deformations of a hyper-elastic body can be derived using Euler-Lagrange equations. This derivation is shorter than the derivation in Chaps. 2, 3, 4, 5, and 6. In fact, most of what is included in Chaps. 2, 3, 4, 5, and 6 is redone here. But the Euler-Lagrange approach requires more challenging mathematics and is included here for those students already familiar with the Euler-Lagrange approach to dynamics.

Euler-Lagrange Equations

Euler-Lagrange equations are the result of setting the variation of the functional J to zero, where J is defined as

$$J = \int_{x_1}^{x_2} f\left(y, \frac{dy}{dx}, x\right) dx. \tag{14.1}$$

Setting the variation of J to zero,

$$\delta J = 0, \tag{14.2}$$

results in the following equations for f,

$$\frac{\partial f}{\partial y} - \frac{d}{dx}\left(\frac{\partial f}{\partial (dy/dx)}\right) = 0 \tag{14.3}$$

Supplementary Information The online version contains supplementary material available at [https://doi.org/10.1007/978-3-031-09157-5_14].

A review of the derivation of Eq. 14.3 is found in Appendix C, along with the extension of this to multiple dimensions.

Define Point Locations Within the Body

The points within a body before deformation are described as $X_i = (X_1, X_2, X_3)$, where X_1, X_2, and X_3 correspond to the initial x, y, and z coordinates of each point, respectively. The corresponding points within the body after a deformation are described as $x_i = (x_1(t), x_2(t), x_3(t))$. Note that the position of each point after deformation is a function of the original location of that point and time, t. That is,

$$
\begin{aligned}
x_1 &= f_1(X_1, X_2, X_3, t) \\
x_2 &= f_2(X_1, X_2, X_3, t) \\
x_3 &= f_3(X_1, X_2, X_3, t).
\end{aligned}
\tag{14.4}
$$

We will assume the body of interest is hyper-elastic. When a hyper-elastic material is deformed, work is done on the material, and energy is stored in the material. As the material is returned to its original shape, the material's energy returns to its original value. The stored energy depends only upon the final position of the points within the body.

Equations of Motion

To describe time-dependent hyper-elastic materials, we need to minimize the Lagrangian of the material. The Lagrangian in classical mechanics is defined as the kinetic energy minus the potential energy: Here these are rescaled by dividing by the initial volume associated with each small region of the material. Thus

$$
L = T - V,
\tag{14.5}
$$

where

L is the Lagrangian per unit initial volume.
T is the kinetic energy per unit initial volume.
V is the potential energy per unit initial volume.

To find the time-dependent motion of the points, classical dynamics dictates that we must minimize J, where J is defined as

$$
J = \int_0^t \int_{V_0} L\left(x_i, \frac{\partial x_i}{\partial x_j}, X_i, t\right) dX_1 \ dX_2 \ dX_3 \ dt.
\tag{14.6}
$$

The kinetic energy of any small region is $\frac{1}{2} mv^2$. Divide this by the initial volume so that

$$T = 1/2 \, \rho_0 \, \dot{x}_i \, \dot{x}_i, \tag{14.7}$$

where

ρ_0 = mass per unit original volume of some small region of the material. (Note that this is *not* the usual definition of density, which is the mass divided by the current volume.)

$$\dot{x}_i = \frac{\partial x_i}{\partial t} = i^{\text{th}} \text{ component of the velocity of the region.} \tag{14.8}$$

We define the total energy per unit initial volume, ϵ_{tot}, as the sum of the deformation energy per unit initial volume, ϵ, and the body energy per unit initial volume, ϵ_{body}, as

$$\epsilon_{\text{tot}} = \epsilon\left(\frac{\partial x_i}{\partial x_j}\right) + \epsilon_{\text{body}}(x_i), \tag{14.9}$$

where the deformation energy per unit initial volume stored in the body, ϵ, is not explicitly a function of time for a hyper-elastic body and

$$\epsilon_{\text{body}}(x_i) = \rho_0 \, g \, x_3 \tag{14.10}$$

for the only body force being gravity in the x_3 direction.

In Eq. 14.9, i and j vary over all values of 1, 2, and 3. The integration is carried out over the entire initial volume, V_0. The term, $\frac{\partial x_i}{\partial X_j}$, is the derivative of each x_i with respect to X_j. i.e.,

$$\frac{\partial x_i}{\partial X_j} \Leftrightarrow \left\{ \frac{\partial x_1}{\partial X_1}, \frac{\partial x_1}{\partial X_2}, \frac{\partial x_1}{\partial X_3}, \frac{\partial x_2}{\partial X_1}, \frac{\partial x_2}{\partial X_2}, \frac{\partial x_2}{\partial X_3}, \frac{\partial x_3}{\partial X_1}, \frac{\partial x_3}{\partial X_2}, \frac{\partial x_3}{\partial X_3} \right\}. \tag{14.11}$$

Requiring $\delta J = 0$ produces the following three Euler-Lagrange Equations:

$$\frac{\partial L}{\partial x_i} - \frac{d}{dX_j}\left(\frac{\partial L}{\partial(\partial x_i/\partial X_j)}\right) - \frac{d}{dt}\left(\frac{\partial L}{\partial \dot{x}_i}\right) = 0, \tag{14.12}$$

giving for $i = 1, 2,$ and 3

$$-\rho_0 \; g \; \delta_{i3} + \frac{d}{dx_j}\left(\frac{\partial \epsilon}{\partial(\partial x_i/\partial x_j)}\right) - \rho_0 \; \ddot{x}_i = 0 \tag{14.13}$$

or

$$\frac{d}{dx_j}\left(\frac{\partial \epsilon}{\partial(\partial x_i/\partial x_j)}\right) - \rho_0 \; g \; \delta_{i3} = \rho_0 \; \ddot{x}_i. \tag{14.14}$$

Equation 14.14 is the equation of motion of the points within the body, i.e., Eq. 4.45.

Equation 14.14 gives the final position of the points within the body once $\epsilon\left(\frac{\partial x_i}{\partial X_j}\right)$ and the boundary conditions are defined. The energy of deformation can be found using the experimental techniques in Chap. 10. The boundary conditions may be Neumann or Dirichlet. Dirichlet boundary conditions just state the final positions of points within the body. Neumann boundary conditions are defined by the applied forces on the body. There are two ways of including the Neumann boundary conditions in these equations. Both methods of including Neumann boundary conditions will be discussed later in this chapter in the section on "Force boundary conditions in simulations". But first let's define the force per unit initial area and the quasi-static equations.

Force

We can identify Eq. 14.14 as Newton's law applied to any small region of the material, divided by the initial volume of this region. That is,

$$\frac{1}{V_0}\Sigma F = \frac{1}{V_0}\;m\;a. \tag{14.15}$$

From Eq. 14.15 we can identify $\frac{d}{dX_j}\left(\frac{\partial \epsilon}{\partial(\partial x_i/\partial X_j)}\right)$ as the force per unit initial volume, the term $\rho_0 \; g \; \delta_{i3}$ as the force of gravity per unit initial volume, and $\rho_0 \; \ddot{x}_i$ as the mass times the acceleration per unit initial volume. The total force of gravity on the body is $\int \rho_0 \; g \; \delta_{i3} dV_0$ and the total boundary force is

$$F_j = \int \frac{\partial}{\partial x_i}\left(\frac{\partial \epsilon}{\partial(\partial x_j/\partial x_i)}\right) dV_0. \tag{14.16}$$

where the integral is over the initial volume of the material. Applying the n-dimensional divergence theorem,

$$\int\int\int (\vec{\nabla} \circ \vec{F})dV_0 = \oint\oint (\vec{F}.\hat{n})dA_0, \qquad (14.17)$$

to Eq. 14.16 gives

$$\int_{V_0} \frac{d}{dX_j}\left(\frac{\partial\epsilon}{\partial(\partial x_i/\partial X_j)}\right)dX_1 \ dX_2 \ dX_3 = \int\left(\frac{\partial\epsilon}{\partial(\partial x_i/\partial X_j)}n_j\right)dA_0, \qquad (14.18)$$

where the integral is over the initial surface area of the body. This gives the force on the body as

$$F_i = \int\left(\frac{\partial\epsilon}{\partial(\partial x_i/\partial x_j)}n_j\right)dA_0, \qquad (14.19)$$

where n_j is the unit vector of the area dA_0 with n_j pointing away from the material region. We can now identify the force per unit initial area on each part of the body as

$$\mathcal{P}_{ij} = \frac{\partial\epsilon}{\partial(\partial x_i/\partial x_j)}. \qquad (14.20)$$

Quasi-static Deformations

A quasi-static deformation is a slow deformation where acceleration is negligible. All that is required is that the acceleration of the material be set to zero in Eq. 14.14, giving

$$0 = \frac{d}{dX_j}\left(\frac{\partial\epsilon}{\partial(\partial x_i/\partial X_j)}\right) - \rho_0 \ g \ \delta_{i3}. \qquad (14.21)$$

Substituting the body force of gravity ($\rho_0 \ g$) into Eq. 14.21 gives

$$\frac{d}{dX_1}\left(\frac{\partial\epsilon}{\partial(\partial x_1/\partial X_1)}\right) + \frac{d}{dX_2}\left(\frac{\partial\epsilon}{\partial(\partial x_1/\partial X_2)}\right) + \frac{d}{dX_3}\left(\frac{\partial\epsilon}{\partial(\partial x_1/\partial X_3)}\right) = 0$$

$$\frac{d}{dX_1}\left(\frac{\partial\epsilon}{\partial(\partial x_2/\partial X_1)}\right) + \frac{d}{dX_2}\left(\frac{\partial\epsilon}{\partial(\partial x_2/\partial X_2)}\right) + \frac{d}{dX_3}\left(\frac{\partial\epsilon}{\partial(\partial x_2/\partial X_3)}\right) = 0$$

$$\frac{d}{dX_1}\left(\frac{\partial\epsilon}{\partial(\partial x_3/\partial X_1)}\right) + \frac{d}{dX_2}\left(\frac{\partial\epsilon}{\partial(\partial x_3/\partial X_2)}\right) + \frac{d}{dX_3}\left(\frac{\partial\epsilon}{\partial(\partial x_3/\partial X_3)}\right) - \rho_0 \ g \ = 0.$$

$$(14.22)$$

These are Eq. 4.45 with a_j zero.

Force Boundary Conditions in Simulations

There are two ways to include force boundary conditions in simulations. Consider a boundary region, Fig. 14.1. The nine terms in $\frac{d}{dX_j}\left(\frac{\partial \epsilon}{\partial(\partial x_i/\partial X_j)}\right)$ contain the forces on a portion of the material from the surrounding regions. For a boundary node, there are no regions adjacent to the boundary region of the node (i.e., the red surface in Fig. 14.1). The external force on this region can be added either by substituting the boundary force directly in the place of the boundary terms $\frac{\partial \epsilon}{\partial(\partial x_i/\partial X_j)}$, or alternatively the boundary force can be included by adding the energy stored in the material due to the work done by the boundary force. The stored energy is calculated as the negative of the work done by the bounding force,

$$\epsilon_{\text{bound}} = -\int_{\text{initial } \vec{x}}^{\text{final } \vec{x}} \vec{F} \circ d\vec{x}. \tag{14.23}$$

This latter method was used for the simulations in Chaps. 7, 8, 9, 10, 11, 12, 13, and 14.

Fig. 14.1 Small region in space with one exposed face (red) to which a boundary force is applied

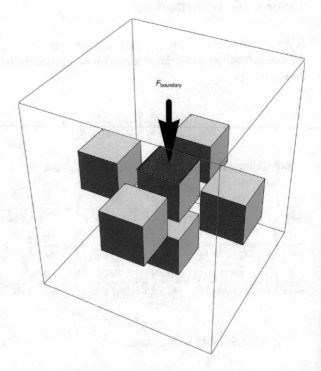

Discrete Energy Equations Used in Numerical Simulations

In numerical simulations we choose to apply Neumann boundary conditions by adding the energy associated with the work done by the external forces on the boundary nodes. Thus for quasi-static deformations,

$$\epsilon_{\text{tot}} = \epsilon\left(\frac{\partial x_i}{\partial X_j}\right) + \epsilon_{\text{body}}(x_i) + \epsilon_{\text{boundaryforces}}, \qquad (14.24)$$

where

$$\epsilon_{\text{boundary forces}} = -\text{work done by boundary forces} = -\int_{\vec{X}}^{\vec{x}} F_{\text{boundary}} \circ d\vec{x}$$

$$= -\int_{\vec{X}}^{\vec{x}} f_{\text{boundary}} \circ A_0 \, d\vec{x}, \qquad (14.25)$$

where f_{boundary} = force per unit initial area of the boundary and A_0 = initial area that the force is applied against. (Note that this is *not* Cauchy stress, which is the force divided by the current area.)

The quantity to be minimized is

$$J = \int_{V_0} \epsilon_{\text{tot}} dX_1 dX_2 dX_3. \qquad (14.26)$$

To numerically simulate Eq. 14.26, we divide the body into N_v regions defined by N_n nodes. The nodes are considered to be close together and many in number so that the integral can be approximated as a sum. The path taken to the final position is taken in small steps so that the force may be considered constant over each step. Then Eq. 14.26 becomes

$$J = \sum_{i=1}^{N_v} \epsilon\left(\frac{\partial x_i}{\partial X_j}\right) V_i + \sum_{i=1}^{N_n} \epsilon_{\text{body}}(x_i) V_i + \sum_{j=1}^{N_b} \vec{f}_{\text{app}} \circ \left(\vec{x}_j - \vec{X}_j\right) A_j, \qquad (14.27)$$

where

N_b = number of boudary nodes with applied boundary forces
N_v = number of volumes
N_n = number of nodes
A_j = inial area of node j.
V_i = initial volume of node i.

Noting that $\sum_{i=1}^{N_b} \vec{f}_{\text{app}} \circ \vec{X}_j A_j$ is a constant for each step of the simulation, Eq. 14.27 can be simplified to

$$J = \sum_{i=1}^{N_v} \epsilon_i V_i + \sum_{k=1}^{N_n} \rho_0 g x_k \overline{V}_k - \sum_{j=1}^{N_b} \overrightarrow{f}_{app} \circ \overrightarrow{x}_j A_j, \qquad (14.28)$$

which is Eq. 6.30.

With Eqs. 14.28 discretized simulations can proceed with finite difference, finite element, Rayleigh-Ritz, or other numerical technique. In the examples in Chaps. 7, 8, 9, 10, 11, and 12, a simple energy minimization is used.

Energy per Unit Initial Volume

Now we need to define the energy per unit original volume, $\epsilon\left(\frac{\partial x_i}{\partial X_j}\right)$. For an isotropic body, rotation of the body before or after the deformation will not change the energy of the body, which is associated with its deformation. Unfortunately, the individual values of $\frac{\partial x_i}{\partial X_j}$ do change with these rotations. Thus, not any function of $\frac{\partial x_i}{\partial X_j}$ is permitted for an isotropic body. For an isotropic body, the following combinations of $\frac{\partial x_i}{\partial X_j}$ do not change with these rotations:

$$
\begin{aligned}
\mathcal{I}_1 &= \vec{a} \cdot \vec{a} + \vec{b} \cdot \vec{b} + \vec{c} \cdot \vec{c} \\
\mathcal{I}_2 &= \left(\vec{a} \times \vec{b}\right) \cdot \left(\vec{a} \times \vec{b}\right) + \left(\vec{a} \times \vec{c}\right) \cdot \left(\vec{a} \times \vec{c}\right) + \left(\vec{b} \times \vec{c}\right) \cdot \left(\vec{b} \times \vec{c}\right) \\
\mathcal{I}_3 &= \vec{a} \cdot \left(\vec{b} \times \vec{c}\right).
\end{aligned}
$$

$$(14.29)$$

These are the invariants of an isotropic body. That these are invariants of $\mathcal{F}$ can be shown by direct algebraic calculation (see Problems 6 and 7). We recognize that for isotropic bodies, ϵ is

$$\epsilon = f(\mathcal{I}_1, \mathcal{I}_2, \mathcal{I}_3). \qquad (14.30)$$

The exact function of ϵ in terms of the I_i values must be determined by experiment. Chapter 10 shows one way this can be accomplished. For anisotropic bodies, the energy of deformation must be invariant to rotations of the body only *after* the deformation. A description of ϵ and how to measure these are found in Chap. 12.

Problems

Problem 1 Derive the 1D equation of motion of a quasi-static deformation without gravity by finding the Euler-Lagrange equations resulting from $\delta J = \delta \int_{X_1}^{X_2} \mathcal{L}\left(x, \frac{\partial x}{\partial X}, X\right) dX = 0$.

Problem 2 Assume a body of mass m suspended by an elastic spring. Limit the motion of the mass to a vertical plane. The kinetic energy is then $T = \frac{1}{2}m\left(\dot{r}^2 + r^2\dot{\theta}^2\right)$ and the potential energy, $V = \frac{1}{2}k(r - r_0)^2 - m\ r\ g\cos\theta$. Compute the Lagrangian and the two equations of motion. (Hint: see Appendix C and use r for x and θ for y.)

Problem 3 Assume the deformation of a uniform rectangular prism of material with length $= 2$ m, width $= 1$ m, and height $= 6$ m can be treated as a 1D deformation as in Problem 1 with $\epsilon = 4N/m^3\left(\frac{dx}{dX} - 1\right)^2$. Apply an extensional force of 2 N to the 2 m end of the prism. Assume no constraints on the lateral expansion of the bar. Use the results of Problem 1 to find how much the bar will expand.

Problem 4 Derive Eq. 14.14 from Eq. 14.6. (Hint. Follow the pattern set in Appendix C.)

Problem 5 Suppose $\epsilon = 2\frac{I}{m^3}\mathcal{I}_1 + 3\frac{I}{m^3}\left(\mathcal{I}_3 - 1\right)^2$. If the deformation is defined as

$$\begin{aligned} x_1 &= 2X_1 - 3X_2 + 2 \\ x_2 &= 4X_2 \\ x_3 &= 2X_1X_2X_3 \end{aligned}$$

What is the force against the small cross-sectional area 0.001 m^2 perpendicular to the x axis?

Problem 6 Show that $\mathcal{I}_1$, $\mathcal{I}_2$, and $\mathcal{I}_3$ are invariant under a general rotation before the deformation $\frac{\partial x_i}{\partial X_j}$. To do this, define $\mathcal{R} = \mathcal{R}_x\mathcal{R}_y\mathcal{R}_z$. Define $\mathcal{F}_{ij} = \frac{\partial x_i}{\partial X_j}$. Calculate $\mathcal{I}_1$, $\mathcal{I}_2$, and $\mathcal{I}_3$ for $\mathcal{F}_{ij}$. Then calculate $\mathcal{I}_1$, $\mathcal{I}_2$, and $\mathcal{I}_3$ for $\mathcal{F}'_{ij}$, where $\mathcal{F}'_{ij} = \frac{\partial x_i}{\partial X_k}\mathcal{R}_{kj}$. Show that the $\mathcal{I}_i$ values are unchanged by the rotation. If an algebraic program is not available, show this result for $\theta_x = 20°$, $\theta_y = 25°$, and $\theta_z = 15°$.

Problem 7 Show that $\mathcal{I}_1$, $\mathcal{I}_2$, and $\mathcal{I}_3$ are invariant under a general rotation after the deformation $\frac{\partial x_i}{\partial X_j}$. To do this, define $\mathcal{R} = \mathcal{R}_x\mathcal{R}_y\mathcal{R}_z$. Define $\mathcal{F}_{ij} = \frac{\partial x_i}{\partial X_j}$. Calculate $\mathcal{I}_1$, $\mathcal{I}_2$, and $\mathcal{I}_3$ for $\mathcal{F}_{ij}$. Then calculate $\mathcal{I}_1$, $\mathcal{I}_2$, and $\mathcal{I}_3$ for $\mathcal{F}'_{ij}$, where $\mathcal{F}'_{ij} = \mathcal{R}_{ik}\frac{\partial x_k}{\partial x_j}$. Show that the $\mathcal{I}_i$ values are unchanged by the rotation. If an algebraic program is not available, show this result for $\theta_x = 20°$, $\theta_y = 25°$, and $\theta_z = 15°$.

Problem 8 Write out the three equations of motion (Eq. 14.14) in terms of $\frac{\partial x_i}{\partial X_j}$ if $\epsilon = \mathcal{I}_1$.

Chapter 15
Linear Elasticity

This chapter is provided for students who are already familiar with existing theories for linear elasticity. A Taylor's expansion of ϵ in terms of $\frac{\partial x_i}{\partial X_j}$ produces the usual equations of linear elasticity keeping only the lowest order terms. An infinitesimal model is proposed for small deformations and large angular rotations.

Comparison to Linear Elasticity

This section will compare the equations presented in this text to those of linear elasticity. The linear elastic equations provide a linear relationship between stress and strain. The theory of linear elasticity is limited to small deformations and small rotations. For small deformations and small rotations, the stress in linear elasticity is the same as engineering stress:

$$\sigma_{ij} = \mathcal{P}_{ij} \text{ (for small deformations)} \tag{15.1}$$

since the initial and final areas are essentially the same,

Linear strain, ε, is defined as a symmetric matrix and is related to the deformation gradient tensor, $\mathcal{F}$, as

$$\varepsilon = 1/2\left(\mathcal{F} + \mathcal{F}^T\right) \tag{15.2}$$

or

$$\varepsilon_{ij} = 1/2\left(\frac{\partial x_i}{\partial X_j} + \frac{\partial x_j}{\partial X_i}\right). \tag{15.3}$$

© The Author(s), under exclusive license to Springer Nature Switzerland AG 2022
H. Hardy, *Engineering Elasticity*, https://doi.org/10.1007/978-3-031-09157-5_15

Notational warning: Note that ϵ is the energy of deformation, but ε_{ij} is the linear strain matrix.

In linear elasticity theory, stress is usually expressed directly in terms of strain, e.g.,

$$\sigma_{ij} = \frac{\nu Y}{(1+\nu)(1-2\nu)}(\varepsilon_{11} + \varepsilon_{22} + \varepsilon_{33})\delta_{ij} + \left(\frac{Y}{(1+\nu)}\right)\varepsilon_{ij}. \qquad (15.4)$$

Equation 4.41 gives $\mathcal{P}_{ij}$ in terms of energy per unit initial volume, ϵ. To show that Eq. 4.41 is the same as Eq. 15.4 for small deformations and small rotations, it is necessary to expand ϵ in a Taylor series of $\frac{\partial x_i}{\partial X_j}$. This is straightforward, but takes quite a bit of algebra.

Linear Elasticity

We wish to express ϵ in the lowest order, nonzero terms of $\frac{\partial x_i}{\partial X_j}$. To do this, a Taylor expansion of $\epsilon\left(\frac{\partial x_i}{\partial X_j}\right)$ gives

$$\epsilon\left(\frac{\partial x_i}{\partial X_j}\right) = \epsilon\left(\frac{\partial x_i}{\partial X_j}\right)_0 + \sum_{i=1}^{3}\sum_{j=1}^{3}\left(\frac{\partial \epsilon}{\partial(\partial x_i/\partial X_j)}\right)_0\left(\frac{\partial x_i}{\partial X_j} - \left(\frac{\partial x_i}{\partial X_j}\right)_0\right)$$

$$+ \frac{1}{2}\sum_{k=1}^{3}\sum_{m=1}^{3}\sum_{i=1}^{3}\sum_{j=1}^{3}\left(\frac{\partial}{\partial(\partial x_k/\partial X_m)}\frac{\partial \epsilon}{\partial(\partial x_i/\partial X_j)}\right)_0 \qquad (15.5)$$

$$\left(\frac{\partial x_i}{\partial X_j} - \left(\frac{\partial x_i}{\partial X_j}\right)_0\right)\left(\frac{\partial x_k}{\partial X_m} - \left(\frac{\partial x_k}{\partial X_m}\right)_0\right) + \dots.$$

where the subscript "0" indicates to evaluate the term before any deformation has occurred. That is, when

$$\frac{\partial x_i}{\partial X_j} = \delta_{ij}, \text{ where } \delta_{ij} = 1 \text{ if } i = j \text{ and } 0 \text{ if } i \neq j. \qquad (15.6)$$

To expand and simplify Eq. 15.5, we will approach each term separately.

Zeroth-Order Term in Equation 15.5

The first term in Eq. 15.5, $\epsilon\left(\frac{\partial x_i}{\partial X_j}\right)_0$, is a constant. Since ϵ only appears in physics equations that contain the derivatives of ϵ, the first term has no effect on the physics of the problem and can be ignored.

First-Order Term in Equation 15.5

If a material has no pre-stress, the internal forces in the material are zero when no deformation has occurred. That is, $\mathcal{P}_{ij} = 0$, if there has been no deformation. Since this is the case for all points within the body, Eq. 4.41 requires

$$\left(\frac{\partial \epsilon}{\partial (\partial x_i/\partial X_j)}\right)_0 = 0. \tag{15.7}$$

This leaves the second-order term in Eq. 15.5 as the leading term in the Taylor expansion of ϵ. Thus to lowest order in $\frac{\partial x_i}{\partial X_j}$,

$$\epsilon = \frac{1}{2}\sum_{k=1}^{3}\sum_{m=1}^{3}\sum_{i=1}^{3}\sum_{j=1}^{3}\left(\frac{\partial}{\partial(\partial x_k/\partial X_m)}\frac{\partial \epsilon}{\partial(\partial x_i/\partial X_j)}\right)_0\left(\frac{\partial x_i}{\partial X_j} - \left(\frac{\partial x_i}{\partial X_j}\right)_0\right)$$
$$\times\left(\frac{\partial x_k}{\partial X_m} - \left(\frac{\partial x_k}{\partial X_m}\right)_0\right)$$

$$\tag{15.8}$$

Before evaluating this term, let's see what constraint Eq. 15.7 provides. Since ϵ is a function of $\mathcal{I}_1$, $\mathcal{I}_2$, and $\mathcal{I}_3$ for isotropic bodies, Eq. 15.7 can be expanded in terms of the three $\mathcal{I}_i$ invariants. That is,

$$\left(\frac{\partial \epsilon}{\partial(\partial x_i/\partial X_j)}\right)_0 = \left(\frac{\partial \epsilon}{\partial \mathcal{I}_r}\frac{\partial \mathcal{I}_r}{\partial(\partial x_i/\partial X_j)}\right)_0$$
$$= \left(\frac{\partial \epsilon}{\partial \mathcal{I}_1}\frac{\partial \mathcal{I}_1}{\partial(\partial x_i/\partial X_j)}\right)_0 + \left(\frac{\partial \epsilon}{\partial \mathcal{I}_2}\frac{\partial \mathcal{I}_2}{\partial(\partial x_i/\partial X_j)}\right)_0$$
$$+ \left(\frac{\partial \epsilon}{\partial \mathcal{I}_3}\frac{\partial \mathcal{I}_3}{\partial(\partial x_i/\partial X_j)}\right)_0. \tag{15.9}$$

Now substitute the $\mathcal{I}_i$ values in terms of $\frac{\partial x_i}{\partial X_j}$. For example,

$$I_1 = \left(\frac{\partial x_1}{\partial X_1}\right)^2 + \left(\frac{\partial x_2}{\partial X_1}\right)^2 + \left(\frac{\partial x_3}{\partial X_1}\right)^2 + \left(\frac{\partial x_1}{\partial X_2}\right)^2 + \left(\frac{\partial x_2}{\partial X_2}\right)^2$$
$$+ \left(\frac{\partial x_3}{\partial X_2}\right)^2 + \left(\frac{\partial x_1}{\partial X_3}\right)^2 + \left(\frac{\partial x_2}{\partial X_3}\right)^2 + \left(\frac{\partial x_3}{\partial X_3}\right)^2 \tag{15.10}$$

gives

$$\left(\frac{\partial I_1}{\partial(\partial x_i/\partial X_j)}\right)_0 = 2\left(\frac{\partial x_i}{\partial X_j}\right)_0 = 2\delta_{ij} \tag{15.11}$$

and the first term in 15.9 is

$$\left(\frac{\partial \epsilon}{\partial I_1}\frac{\partial I_1}{\partial(\partial x_i/\partial X_j)}\right)_0 = \left(\frac{\partial \epsilon}{\partial I_1}\right)_0\left(\frac{\partial I_1}{\partial(\partial x_i/\partial X_j)}\right)_0 = 2\left(\frac{\partial \epsilon}{\partial I_1}\right)_0\delta_{ij}. \tag{15.12}$$

The expansion of $\left(\frac{\partial I_2}{\partial(\partial x_i/\partial X_j)}\right)_0$ and $\left(\frac{\partial I_3}{\partial(\partial x_i/\partial X_j)}\right)_0$ are straightforward, but a bit messy (see Problems 4 and 5 at the end of this chapter). The results substituted into Eq. 15.9 gives

$$\left(\frac{\partial \epsilon}{\partial(\partial x_i/\partial X_j)}\right)_0 = \left(2\frac{\partial \epsilon}{\partial I_1} + 4\frac{\partial \epsilon}{\partial I_2} + \frac{\partial \epsilon}{\partial I_3}\right)_0\delta_{ij}. \tag{15.13}$$

Thus, if the material is not pre-stressed,

$$\left(2\frac{\partial \epsilon}{\partial I_1} + 4\frac{\partial \epsilon}{\partial I_2} + \frac{\partial \epsilon}{\partial I_3}\right)_0 = 0 \tag{15.14}$$

or

$$\left(\frac{\partial \epsilon}{\partial I_3}\right)_0 = -2\left(\frac{\partial \epsilon}{\partial I_1}\right)_0 - 4\left(\frac{\partial \epsilon}{\partial I_2}\right)_0 \tag{15.15}$$

or

$$\left(\frac{\partial \epsilon}{\partial I_2}\right)_0 = -\frac{1}{4}\left(2\left(\frac{\partial \epsilon}{\partial I_1}\right)_0 + \left(\frac{\partial \epsilon}{\partial I_3}\right)_0\right). \tag{15.16}$$

Second-Order Term in Equation 15.5

The evaluation of Eq. 15.8 remains, subject to the constraint given in Eq. 15.14. As before, expand ϵ in terms of I_r invariants.

The center term

$$
\left(\frac{\partial}{\partial(\partial x_k/\partial X_m)} \frac{\partial \epsilon}{\partial(\partial x_i/\partial X_j)} \right)_0 = \left(\frac{\partial^2 \epsilon}{\partial I_r \partial I_s} \left(\frac{\partial I_s}{\partial(\partial x_i/\partial X_j)} \right) \left(\frac{\partial I_r}{\partial(\partial x_k/\partial X_l)} \right) \right)_0
$$
$$
+ \left(\frac{\partial \epsilon}{\partial I_r} \left(\frac{\partial^2 I_r}{\partial(\partial x_i/\partial X_j)\partial(\partial x_k/\partial X_l)} \right) \right)_0,
$$

$$(15.17)$$

which gives for Eq. 15.8 the incredibly messy result,

$$
\epsilon = \frac{1}{2} \sum_{k=1}^{3} \sum_{m=1}^{3} \sum_{i=1}^{3} \sum_{j=1}^{3} \left(\sum_{s=1}^{3} \sum_{r=1}^{3} \frac{\partial^2 \epsilon}{\partial I_r \partial I_s} \left(\frac{\partial I_s}{\partial(\partial x_i/\partial X_j)} \right) \left(\frac{\partial I_r}{\partial(\partial x_k/\partial X_l)} \right) \right)
$$
$$
\left(\left(\frac{\partial x_i}{\partial X_j} \right) - \left(\frac{\partial x_i}{\partial X_j} \right)_0 \right) \left(\left(\frac{\partial x_k}{\partial X_m} \right) - \left(\frac{\partial x_k}{\partial X_m} \right)_0 \right) + \frac{1}{2} \sum_{k=1}^{3} \sum_{m=1}^{3} \sum_{i=1}^{3} \sum_{j=1}^{3}
$$
$$
\left(\sum_{r=1}^{3} \frac{\partial \epsilon}{\partial I_r} \left(\frac{\partial^2 I_r}{\partial(\partial x_i/\partial X_j)\partial(\partial x_k/\partial X_l)} \right) \right)_0
$$
$$
\left(\left(\frac{\partial x_i}{\partial X_j} \right) - \left(\frac{\partial x_i}{\partial X_j} \right)_0 \right) \left(\left(\frac{\partial x_k}{\partial X_m} \right) - \left(\frac{\partial x_k}{\partial X_m} \right)_0 \right)
$$

$$(15.18)$$

with the conditions given in Eqs. 15.6 and 15.14 and ϵ a continuous function of I_i so that

$$
\frac{\partial \epsilon}{\partial I_r \partial I_s} = \frac{\partial \epsilon}{\partial I_s \partial I_r}. \tag{15.19}
$$

Equation 15.18 has $3^6 = 729$ terms in the first sum, s_1, and $3^5 = 243$ terms in the second sum, s_2. An algebraic solver is used to simplify Eq. 15.18 (see Problem 10). With the help of the solver, we find that the first sum in Eq. 15.18 reduces to

$$
s_1 = \frac{1}{2} \left(4 \frac{\partial^2 \epsilon}{\partial I_1^2} + 16 \frac{\partial^2 \epsilon}{\partial I_1 \partial I_2} + 4 \frac{\partial^2 \epsilon}{\partial I_1 \partial I_3} + 16 \frac{\partial^2 \epsilon}{\partial I_2^2} + 8 \frac{\partial^2 \epsilon}{\partial I_2 \partial I_3} + \frac{\partial^2 \epsilon}{\partial I_3^2} \right) \bigg|_0
$$
$$
\left(\frac{\partial x_1}{\partial X_1} + \frac{\partial x_2}{\partial X_2} + \frac{\partial x_3}{\partial X_3} - 3 \right)^2.
$$

$$(15.20)$$

The second sum is a bit messier. Here we expand the second sum, s_2, and add the constraint Eq. 15.14. The result is

$$s_2 = \frac{\partial \epsilon}{\partial \mathcal{I}_1}\left(-3 + \left(\frac{\partial x_1}{\partial X_1}\right)^2 + \left(\frac{\partial x_1}{\partial X_2}\right)^2 + \left(\frac{\partial x_1}{\partial X_3}\right)^2 + \left(\frac{\partial x_2}{\partial X_1}\right)^2 + \left(\frac{\partial x_2}{\partial X_2}\right)^2 \right.$$

$$+ \left(\frac{\partial x_2}{\partial X_3}\right)^2 + \left(\frac{\partial x_3}{\partial X_1}\right)^2 + \left(\frac{\partial x_3}{\partial X_2}\right)^2 + \left(\frac{\partial x_3}{\partial X_3}\right)^2 + 2\frac{\partial x_1}{\partial X_1} + 2\frac{\partial x_2}{\partial X_2} + 2\frac{\partial x_3}{\partial X_3}$$

$$+ 2\frac{\partial x_1}{\partial X_2}\frac{\partial x_2}{\partial X_1} + 2\frac{\partial x_1}{\partial X_3}\frac{\partial x_3}{\partial X_1} + 2\frac{\partial x_2}{\partial X_3}\frac{\partial x_3}{\partial X_2} - 2\frac{\partial x_2}{\partial X_2}\frac{\partial x_3}{\partial X_3} - 2\frac{\partial x_1}{\partial X_1}\frac{\partial x_2}{\partial X_2} - 2\frac{\partial x_1}{\partial X_1}\frac{\partial x_3}{\partial X_3} \right)$$

$$+ \frac{\partial \epsilon}{\partial \mathcal{I}_2}\left(6 + 2\left(\frac{\partial x_1}{\partial X_1}\right)^2 + \left(\frac{\partial x_1}{\partial X_2}\right)^2 + \left(\frac{\partial x_1}{\partial X_3}\right)^2 + \left(\frac{\partial x_2}{\partial X_1}\right)^2 + 2\left(\frac{\partial x_2}{\partial X_2}\right)^2 \right.$$

$$+ \left(\frac{\partial x_2}{\partial X_3}\right)^2 + \left(\frac{\partial x_3}{\partial X_1}\right)^2 + \left(\frac{\partial x_3}{\partial X_2}\right)^2 + 2\left(\frac{\partial x_3}{\partial X_3}\right)^2 - 4\frac{\partial x_1}{\partial X_1} - 4\frac{\partial x_2}{\partial X_2}$$

$$\left. - 4\frac{\partial x_3}{\partial X_3} + 2\frac{\partial x_1}{\partial X_2}\frac{\partial x_2}{\partial X_1} + 2\frac{\partial x_1}{\partial X_3}\frac{\partial x_3}{\partial X_1} + 2\frac{\partial x_2}{\partial X_3}\frac{\partial x_3}{\partial X_2} \right).$$

$$(15.21)$$

Before combining this, notice that the term multiplying $\frac{\partial \epsilon}{\partial \mathcal{I}_1}$ in Eq. 15.21 can be made the same as the term multiplying $\frac{\partial \epsilon}{\partial \mathcal{I}_2}$ in s_2 if $\frac{\partial \epsilon}{\partial \mathcal{I}_1}\left(\frac{\partial x_1}{\partial X_1} + \frac{\partial x_2}{\partial X_2} + \frac{\partial x_3}{\partial X_3} - 3\right)^2$ is added to Eq. 15.21. But of course, adding $\frac{\partial \epsilon}{\partial \mathcal{I}_1}\left(\frac{\partial x_1}{\partial X_1} + \frac{\partial x_2}{\partial X_2} + \frac{\partial x_3}{\partial X_3} - 3\right)^2$ to s_2 in Eq. 15.21 means the same term must also be subtract from s_1 in Eq. 15.20. Finally combining the new s_1 and the new s_2 gives finally (Problem 10)

$$\epsilon = \lambda/2\left(\frac{\partial x_1}{\partial X_1} + \frac{\partial x_2}{\partial X_2} + \frac{\partial x_3}{\partial X_3} - 3\right)^2 + \mu\left(3 - 2\frac{\partial x_1}{\partial X_1} + \left(\frac{\partial x_1}{\partial X_1}\right)^2 + \frac{1}{2}\left(\frac{\partial x_1}{\partial X_2}\right)^2 + \frac{1}{2}\left(\frac{\partial x_1}{\partial X_3}\right)^2 + \frac{\partial x_1}{\partial X_2}\frac{\partial x_2}{\partial X_1}\right.$$

$$+ \frac{1}{2}\left(\frac{\partial x_2}{\partial X_1}\right)^2 - 2\frac{\partial x_2}{\partial X_2} + \left(\frac{\partial x_2}{\partial X_2}\right)^2 + \frac{1}{2}\left(\frac{\partial x_2}{\partial X_3}\right)^2 + \frac{\partial x_1}{\partial X_3}\frac{\partial x_3}{\partial X_1} + \frac{1}{2}\left(\frac{\partial x_3}{\partial X_1}\right)^2 + \frac{\partial x_2}{\partial X_3}\frac{\partial x_3}{\partial X_2} + \frac{1}{2}\left(\frac{\partial x_3}{\partial X_2}\right)^2$$

$$\left. - 2\frac{\partial x_3}{\partial X_3} + \left(\frac{\partial x_3}{\partial X_3}\right)^2\right),$$

$$(15.22)$$

where λ and μ are called the Lame parameters. λ and μ are defined as

$$\lambda = \left(4\frac{\partial^2 \epsilon}{\partial \mathcal{I}_1^2} + 16\frac{\partial^2 \epsilon}{\partial \mathcal{I}_1 \partial \mathcal{I}_2} + 4\frac{\partial^2 \epsilon}{\partial \mathcal{I}_1 \partial \mathcal{I}_3} + 16\frac{\partial^2 \epsilon}{\partial \mathcal{I}_2^2} + 8\frac{\partial^2 \epsilon}{\partial \mathcal{I}_2 \partial \mathcal{I}_3} + \frac{\partial^2 \epsilon}{\partial \mathcal{I}_3^2} - 2\frac{\partial \epsilon}{\partial \mathcal{I}_1}\right)\bigg|_0$$

$$(15.23)$$

and

$$\mu = 2\left(\frac{\partial \epsilon}{\partial \mathcal{I}_1} + \frac{\partial \epsilon}{\partial \mathcal{I}_2}\right)\bigg|_0. \tag{15.24}$$

This complicated result does not look too appealing, but the next section will show that it is equivalent to the standard linear elastic energy equation.

Comparison to Standard Linear Elasticity Equations

In the last section, a Taylor expansion on ϵ found that the leading term for non-pre-stressed materials was expressed by Eq. 15.22. This section shows that this energy function is the same as the standard linear elasticity energy equation as expressed by Landau (*Theory of Elasticity*, 3rd Edition, pg 9, published by Butterworth and Heinemann, 1999). To do this, a change of variables is needed.

Define the displacement vector

$$u_i = x_i - X_i \tag{15.25}$$

so that

$$\frac{\partial u_i}{\partial X_j} = \frac{\partial x_i}{\partial X_j} - \delta_{ij}. \tag{15.26}$$

Define the infinitesimal strain as the following symmetric matrix:

$$\varepsilon_{ij} = \frac{1}{2}\left(\frac{\partial u_i}{\partial X_j} + \frac{\partial u_j}{\partial X_i}\right). \tag{15.27}$$

Using these new variables, Eq. 15.22 becomes

$$\epsilon = \lambda/2(\varepsilon_{11} + \varepsilon_{22} + \varepsilon_{33})^2$$
$$+ \mu\left(\varepsilon_{11}^2 + \varepsilon_{12}^2 + \varepsilon_{13}^2 + \varepsilon_{21}^2 + \varepsilon_{22}^2 + \varepsilon_{23}^2 + \varepsilon_{31}^2 + \varepsilon_{32}^2 + \varepsilon_{33}^2\right). \tag{15.28}$$

Landau, in *Theory of Elasticity*, writes Eq. 15.28 as

$$\epsilon = \epsilon_0 + \frac{1}{2}\lambda\varepsilon_{ii}^2 + \mu\varepsilon_{ik}^2, \tag{15.29}$$

where

ϵ = energy (per unit volume) of linear deformation.

$$\varepsilon_{ii}^2 = \left(\sum\nolimits_{i=1}^{3}\varepsilon_{ii}\right)^2 = (\varepsilon_{11} + \varepsilon_{22} + \varepsilon_{33})^2 \tag{15.30}$$

and

$$\varepsilon_{ik}^2 = \sum\nolimits_{i=1}^{3}\sum\nolimits_{j=1}^{3}\left(\varepsilon_{ij}\right)^2$$
$$= \left(\varepsilon_{11}^2 + \varepsilon_{12}^2 + \varepsilon_{13}^2 + \varepsilon_{21}^2 + \varepsilon_{22}^2 + \varepsilon_{23}^2 + \varepsilon_{31}^2 + \varepsilon_{32}^2 + \varepsilon_{33}^2\right). \tag{15.31}$$

Since ϵ_0 is a constant and has no impact on the physics of the deformation, Eqs. 15.28 and 15.29 are equivalent.

Landau's Energy Equation Leads to Linear Elastic Force-Displacement Equations

Most engineers dealing with linear elasticity are more familiar with the elastic equations expressed in terms of stress and strain instead of in terms of energy as given by Eq. 15.29. This section will show (as Landau did) that Eq. 15.29 leads to the familiar linear stress-strain relationships of linear elasticity.

From Eq. 15.29,

$$\frac{\partial\epsilon}{\partial\left(\partial x_i/\partial X_j\right)} = \lambda(\varepsilon_{11} + \varepsilon_{22} + \varepsilon_{33})\delta_{ij} + 2\mu\varepsilon_{ij}, \tag{15.32}$$

which according to Eq. 4.41 is $\mathcal{P}_{ij}$, i.e.,

$$\mathcal{P}_{ij} = \lambda(\varepsilon_{11} + \varepsilon_{22} + \varepsilon_{33})\delta_{ij} + 2\mu\varepsilon_{ij}. \tag{15.33}$$

Define Young's modulus, Y, and Poisson's ratio, ν, in terms of the Lame parameters, λ and μ, as follows:

$$\lambda = \frac{\nu Y}{(1+\nu)(1-2\nu)} \tag{15.34}$$

and

$$\mu = \frac{Y}{2(1+\nu)}. \tag{15.35}$$

Equation 15.33 becomes

$$\mathcal{P}_{ij} = \frac{\nu Y}{(1+\nu)(1-2\nu)}(\varepsilon_{11} + \varepsilon_{22} + \varepsilon_{33})\delta_{ij} + \frac{Y}{(1+\nu)}\varepsilon_{ij}. \tag{15.36}$$

From Eq. 15.1, we have

$$\sigma_{ij} = \mathcal{P}_{ij}. \tag{15.37}$$

Equation 15.36 can now be written as

$$\sigma_{ij} = \frac{\nu Y}{(1+\nu)(1-2\nu)}(\varepsilon_{11} + \varepsilon_{22} + \varepsilon_{33})\delta_{ij} + \left(\frac{Y}{(1+\nu)}\right)\varepsilon_{ij}. \tag{15.38}$$

This is usually inverted to give

$$\epsilon_{ij} = \frac{(1+\nu)}{Y}\sigma_{ij} - \frac{\nu}{Y}\sigma_{kk}\delta_{ij} \tag{15.39}$$

or

$$
\begin{aligned}
\varepsilon_{11} &= \frac{1}{Y}(\sigma_{11} - \nu(\sigma_{22} + \sigma_{33}) \\
\varepsilon_{22} &= \frac{1}{Y}(\sigma_{22} - \nu(\sigma_{11} + \sigma_{33}) \\
\varepsilon_{33} &= \frac{1}{Y}(\sigma_{33} - \nu(\sigma_{11} + \sigma_{22}) \\
\varepsilon_{12} &= \frac{1+\nu}{Y}\sigma_{12} \\
\varepsilon_{13} &= \frac{1+\nu}{Y}\sigma_{13} \\
\varepsilon_{23} &= \frac{1+\nu}{Y}\sigma_{23}
\end{aligned}
\tag{15.40}
$$

with

$$
\begin{aligned}
\varepsilon_{21} &= \varepsilon_{12} \\
\varepsilon_{31} &= \varepsilon_{13} \\
\varepsilon_{32} &= \varepsilon_{23}.
\end{aligned}
\tag{15.41}
$$

Equations 15.40 and 15.41 are the most common equations used to describe linear elasticity.

Some Limitations of Linear Elasticity

The linear elastic equations derived in the last section have two conditions on their use.

1. Deformations must be small, so that changes in ϵ_{ij} are less than 5% of the initial energy.
2. Rotations must be small so that angles of rotation are less than 10 degrees.

The first constraint is obvious because only the lowest nonzero term was kept in the Taylor expansion in Eq. 15.5. The need for the second constraint is best illustrated by an example.

Suppose a body is rotated slowly through a large angle, θ, about the z-axis with only infinitesimal forces applied. As a result, the new positions of each point in the body, x_i, in terms of the initial positions, X_i, are as follows:

$$
\begin{aligned}
x_1 &= X_1 \cos\theta - X_2 \sin\theta \\
x_2 &= X_1 \sin\theta + X_2 \cos\theta \\
x_3 &= X_3
\end{aligned}
\tag{15.42}
$$

using Eq. 15.27 to calculate ε_{ij} gives

$$
\varepsilon_{ij} =
\begin{pmatrix}
-(1-\cos\theta) & 0 & 0 \\
0 & -(1-\cos\theta) & 0 \\
0 & 0 & 0
\end{pmatrix},
\tag{15.43}
$$

which means for finite angles the body has undergone a finite deformation as well as being rotated – but only infinitesimal forces were applied, so no deformation should have occurred. Clearly something is amiss; however, expanding the $\cos\theta$ in a Taylor series shows that the errors generated are of the order of θ^2 or higher, so that ε_{ij} can be treated as zero if the angle of rotation is small, and therefore for small angles the deformation is negligible.

Advantage of Linear Elasticity

The main advantage of linear elasticity is that the relationship between stress and strain is linear. This means that numerical linear solvers can be used in computer codes to solve Eq. 15.40. This is much faster than the nonlinear solvers used for finite deformations. Also, as shown earlier in this chapter, Eq. 15.40 can be inverted to give equations of strain as a function of stress.

An interesting side note is that Eq. 15.27 insures that ϵ_{ij} is symmetric (i.e., $\epsilon_{ij} = \epsilon_{ji}$), and because of this Eq. 15.38 requires σ_{ij} to be symmetric. That is, the symmetry of σ_{ij} comes directly out of the Taylor expansion of ϵ. The additional condition which often stated as the "sum of torques (or couples) must be zero" to insure σ_{ij} to be symmetric is not required.

Small Deformations with Rotations

In this section we will relax the second condition of linear elasticity (i.e., that rotations be small). To do this it is necessary to deal with some nonlinearity. One way to obtain a description of infinitesimal elasticity without the rotation problem just encountered is to Taylor expand $\epsilon(\mathcal{I}_i)$ in terms of $\mathcal{I}_i$ instead of in terms of $\frac{\partial x_i}{\partial X_j}$. This will keep the rotational independence of ϵ, because all $\mathcal{I}_i$'s are invariant under rotation. In addition, we would like the form of $\epsilon(\mathcal{I}_i)$ to mimic the linear elastic relationship between stress and strain. That is, it should yield no stress where there is no strain. It should produce a linear relationship between stress and strain for small deformations. Moreover, it should have only two constants which can be expressed in terms of Young's modulus and Poisson's ratio.

The Taylor expansion of ϵ in terms of $\mathcal{I}_i$ is

$$\epsilon(\mathcal{I}_i) = \epsilon(\mathcal{I}_i)|_0 + \frac{\partial \epsilon}{\partial \mathcal{I}_1}\bigg|_0 (\mathcal{I}_1 - (\mathcal{I}_1)_0)$$

$$+ \frac{\partial \epsilon}{\partial \mathcal{I}_2}\bigg|_0 (\mathcal{I}_2 - (\mathcal{I}_2)_0) + \frac{\partial \epsilon}{\partial \mathcal{I}_3}\bigg|_0 (\mathcal{I}_3 - (\mathcal{I}_3)_0) + \ldots \qquad (15.44)$$

or to first order

$$\epsilon(\mathcal{I}_i) = \epsilon(\mathcal{I}_i)|_0 + \frac{\partial \epsilon}{\partial \mathcal{I}_1}\bigg|_0 (\mathcal{I}_1 - 3)$$

$$+ \frac{\partial \epsilon}{\partial \mathcal{I}_2}\bigg|_0 (\mathcal{I}_2 - 3) + \frac{\partial \epsilon}{\partial \mathcal{I}_3}\bigg|_0 (\mathcal{I}_3 - 1). \qquad (15.45)$$

Dropping the nonphysical first term gives

$$\epsilon(\mathcal{I}_i) = \frac{\partial \epsilon}{\partial \mathcal{I}_1}\bigg|_0 (\mathcal{I}_1 - 3) + \frac{\partial \epsilon}{\partial \mathcal{I}_2}\bigg|_0 (\mathcal{I}_2 - 3) + \frac{\partial \epsilon}{\partial \mathcal{I}_3}\bigg|_0 (\mathcal{I}_3 - 1). \qquad (15.46)$$

As before,

$$\frac{\partial \epsilon(\mathcal{I}_i)}{\partial (\partial x_i / \partial X_j)}\bigg|_0 = 0, \qquad (15.47)$$

so that there is no pre-stress. This again leads to the condition given in Eq. 15.16.

To apply this condition, define

$$\left.\frac{\partial \epsilon}{\partial \mathcal{I}_1}\right|_0 = A$$

$$\left.\frac{\partial \epsilon}{\partial \mathcal{I}_2}\right|_0 = B \qquad (15.48)$$

$$\left.\frac{\partial \epsilon}{\partial \mathcal{I}_3}\right|_0 = C$$

and Eq. 15.16 gives

$$B = -(2A - 4C)/4 \qquad (15.49)$$

so that Eq. 15.46 is

$$\epsilon(\mathcal{I}_i) = A(\mathcal{I}_1 - 3) - (2A + 4C)/4(\mathcal{I}_2 - 3) + C(\mathcal{I}_3 - 1). \qquad (15.50)$$

Unfortunately, Eq. 15.50 is unstable so that the next term needs to be included in the Taylor's expansion of ϵ. The ϵ representation needed then is

$$\begin{aligned}
\epsilon(\mathcal{I}_i) = {} & A(\mathcal{I}_1 - 3) - (2A + 4C)/4(\mathcal{I}_2 - 3) + C(\mathcal{I}_3 - 1) + D(\mathcal{I}_1 - 3)^2 \\
& + G(\mathcal{I}_2 - 3)^2 + H(\mathcal{I}_3 - 1)^2 + K(\mathcal{I}_1 - 3)(\mathcal{I}_2 - 3) \\
& + L(\mathcal{I}_1 - 3)(\mathcal{I}_3 - 1) + M(\mathcal{I}_2 - 3)(\mathcal{I}_3 - 1)
\end{aligned}$$

$$(15.51)$$

with

$$D = \left(\frac{\partial^2 \epsilon}{\partial \mathcal{I}_1^2}\right)_0$$

$$G = \left(\frac{\partial^2 \epsilon}{\partial \mathcal{I}_2^2}\right)_0$$

$$H = \left(\frac{\partial^2 \epsilon}{\partial \mathcal{I}_3^2}\right)_0$$

$$K = \left(\frac{\partial^2 \epsilon}{\partial \mathcal{I}_1 \partial \mathcal{I}_2}\right)_0 \qquad (15.52)$$

$$L = \left(\frac{\partial^2 \epsilon}{\partial \mathcal{I}_1 \partial \mathcal{I}_3}\right)_0$$

$$M = \left(\frac{\partial^2 \epsilon}{\partial \mathcal{I}_2 \partial \mathcal{I}_3}\right)_0.$$

We wish to construct an ϵ which has only two parameters that can be related to Y and ν. Equation 15.52 provides too many parameters. As with linear elasticity, we need only two parameters. One way to simplify Eq. 15.51 for infinitesimal deformations is to set all constants in Eq. 15.51 to zero except for A and H. When this is done, Eq. 15.51 for infinitesimal deformations becomes

$$\epsilon(\mathcal{I}_i) = A(\mathcal{I}_1 - 3) - A/2(\mathcal{I}_2 - 3) + H(\mathcal{I}_3 - 1)^2 \qquad (15.53)$$

with

$$A = \frac{Y}{2(1+\nu)}$$

$$H = \frac{(1-\nu)Y}{2(1-2\nu)(1+\nu)}, \qquad (15.54)$$

or dropping constant terms and simplifying

$$\epsilon(\mathcal{I}_i) = \frac{Y}{2(1+\nu)}\left(\left(\mathcal{I}_1 - \frac{1}{2}\mathcal{I}_2\right) + \frac{1-\nu}{(1-2\nu)}(\mathcal{I}_3 - 1)^2\right). \qquad (15.55)$$

Equation 15.55 is an energy for small deformations which is rotationally invariant.

Note that as ν approaches 0.5, as required for incompressibility, the second term in Eq. 15.55 goes to infinity. Setting this second term much greater than the first term is what is used in the simulations in Chaps. 7, 8, 9, 10, 11, 12, 13, and 14 to simulate incompressibility or near incompressibility. Also, A in Eq. 15.54 allows Y to be defined in terms of A and produce the linear fit shown in Fig. 8.4.

Problems

Problem 1 Show that Eq. 15.28 follows from taking the derivative of ϵ in Eq. 15.22 with respect to $\frac{\partial x_i}{\partial X_i}$.

Problem 2 Assume that $\lambda = 83.3$ GPa and $\mu = 0.0167$ GPa. (These are appropriate values for rubber). Using the linear elasticity energy equation, Eq. 15.22, by how much will a rotation of $25°$ about the z axis increase ϵ of a 0.5 m × 0.5 m × 0.5 m cube?

Problem 3 Derive Eq. 15.43 from Eqs. 15.27 and 15.42.

Problem 4 Expand $\left(\dfrac{\partial I_2}{\partial(\partial x_i/\partial X_j)}\right)_0$ and show this can be written as $4\,\delta_{ij}$.

Problem 5 Expand $\left(\dfrac{\partial I_3}{\partial(\partial x_i/\partial X_j)}\right)_0$ and show this can be written as δ_{ij}.

Problem 6 Show that Eq. 15.55 satisfies Eq. 15.14.

Problem 7 Show that Eq. 15.55 satisfies Eq. 15.34 by substituting ϵ into Eq. 15.23.

Problem 8 Show that Eq. 15.55 satisfies Eq. 15.35 by substituting ϵ into Eq. 15.24.

Problem 9 Show that Eq. 15.40 follow from Eq. 15.39.

Problem 10 Show that Eq. 15.40 follow from Eq. 15.38.

Problems 11, 12, and 13 contain the derivation from Eqs. 15.18 to 15.22 in three steps. Problem 11 computes the first sum in 15.18. Problem 12 computes the second sum in 15.18. Problem 13 puts the two sums together to get Eq. 15.28. Since 15.18 has almost 1000 terms, an algebraic solver is required for each step.

Problem 11 Use an algebraic solver program to show that Eq. 15.20 follows from the first sum in Eq. 15.18.

Problem 12 Use an algebraic solver program to show that Eq. 15.21 follows from the second sum in Eq. 15.18.

Problem 13 Show that Eq. 15.22 follows from the sum of Eqs. 15.20 and 15.21.

Chapter 16
Classical Finite Elasticity

In this chapter, we will compare the "standard" equations of finite elasticity (i.e., classical finite elasticity) to those used in this book. This includes a comparison of both "classical finite elasticity" and "rubber elasticity". The first section contrasts classical finite elasticity in the literature to that presented in this text. The second section connects the equations and symbols of the classical finite elasticity text by Spencer to this text. The last section connects the equations and symbols of Rivlin's papers on rubber elasticity to this text.

Classical Elasticity

Each of the authors in classical elasticity has their own unique style in covering this topic; however, there are five basic threads that run through most classical elasticity and continuum mechanics books. These are as follows:

1. Many stress and strain measures are used to describe finite elasticity.
2. Deformation is usually described with a symmetric tensor.
3. Stress is usually described with a symmetric tensor.
4. The equation of motion of points (or particles) within the body is stated in the material coordinates (as well as the reference coordinates).
5. The tensor properties of matrices are used to describe the theory.

This text shows that none of these are necessary to describe finite elasticity for the engineer.

1. Spencer[1], Ogden[2], Bonet[3], Atkin[4], etc., all define more than a dozen stress and strain tensors, and Tables 16.1 and 16.2 list named stress and strain measures used

Supplementary Information The online version contains supplementary material available at [https://doi.org/10.1007/978-3-031-09157-5_16].

229

H. Hardy, *Engineering Elasticity*, https://doi.org/10.1007/978-3-031-09157-5_16

Table 16.1 Sampling of strain measures used by different authors

Strain	Spencer	Ogden	Bonet	Atkin	Hardy
Deformation gradient tensor	F	A	F	F	$\mathcal{F}$
Displacement gradient tensor	$F - I$	–	–	$H = F - I$	–
Left Cauchy-Green deformation tensor	$B = FF^T$	AA^T	$b = FF^T$	$B = FF^T$	–
Right Cauchy-Green deformation tensor	$C = F^T F$	AA^T	$C = F^T F$	$C = F^T F$	–
Lagrangian strain tensor	$\gamma = 1/2(C - I)$	$E = 1/2(A^T A - I)$	$E = 1/2(C - I)$	–	–
Eulerian strain tensor	$\eta = 1/2(I - B^{-1})$	$1/2(I - BB^T)$	$e = 1/2(I - b^{-1})$	–	–
Infinitesimal strain tensor	$E = 1/2(F + F^T) - I$	–	–	$E = 1/2(H + H^T)$	–
Right stretch tensor	$U = R^T F$	$U = R^T A$	$U = R^T F$	$U = R^T F$	–
Left stretch tensor	$V = FR^T$	$V = AR^T$	$V = FR^T$	$V = FR^T$	–
Inverse deformation gradient tensor	–	$B = (A^T)^{-1}$	–	–	–
General Lagrangian strain tensor	–	$F(U) = \sum_{i=1}^{3} f(\lambda_i) u^{(i)} \bigotimes u^{(i)}$	–	–	–
Tensor of vth-order elastic moduli	–	$A^v = \partial^v S / \partial A^v$	–	–	–
Fourth-order tensor	–	$L = \partial F / \partial U$	–	–	–
Sixth-order tensor	–	$L^2 = \partial^2 F / \partial U^2$	–	–	–
Biot strain tensor	–	$F(U) = U - I$	–	–	–
Almasi strain tensor	–	$F(U) = 1/2(I - B^T B)$	–	–	–
Material logarithmic strain tensors	–	–	$E^{(0)} = \ln U$	–	–
Spatial logarithmic strain tensors	–	–	$e^{(0)} = \ln V$	–	–
Strain deviator	–	–	–	$E'_{KL} = E_{KL} - \dfrac{1}{3} E_{MM} \delta_{KL}$	–

Table 16.2 Sampling of stress measures used by different authors

Stress	Spencer	Ogden	Bonet	Atkin	Hardy
Cauchy stress tensor	T	T	σ	T	–
Stress deviatoric tensor	$S = T - \dfrac{1}{3}J_1 I,$ $J_1 = \text{Trace}(T)$	–	–	–	–
Kirchhoff stress tensor	–	$\hat{T} = JT$ $J = |J|$	$\tau = J\sigma$ $J = |J|$	–	–
Nominal stress tensor	–	$S = JB^T T$	–	–	–
First Piola-Kirchhoff stress tensor	$\pi = (|F|)F^{-1}T$	$S^T = JT^T B$	$P = J\sigma F^{-T}$	$S = JF^T T$	–
Second Piola-Kirchhoff stress tensor	$P = \pi(F^{-1})^T$	$T^{(2)} = B^T \hat{T} B$	$S = JF^{-1}\sigma F^{-T}$	$\bar{S}_{KL} = X_{Li}S_{ki}$	–
Biot (or Jaumann) stress tensor	–	$T^{(1)} = 1/2\left(T^{(2)}U + UT^{(2)}\right)$	–	–	–
Conjugate stress tensors	–	$E^{(m)} =$ $\ln U,$ if m $= 0$ $1/m(U^M - I),$ if m $\neq 0$	–	–	–
Purely normal stress	–	–	–	$S = -pI$	–
Engineering stress		–	–		$\mathcal{P}$

by these authors. There are more authors of course, but the trend is the same. Many measures are defined for both stress and strain. As shown in this text, none of the many stress and strain measures defined by these authors are used in this text except the "deformation gradient tensor" and the "transpose of the first Piola-Kirchhoff stress tensor", called engineering stress in this text.

Tables 16.1 and 16.2 show that more than a dozen stress and strain measures are used by each author to describe classical finite elasticity. The column labeled "Hardy" lists the stress and strain measures used in this text. Notice that there is only one stress measure and one strain measure used in this text. None of the dozen or more stress and strain measures used by other authors is needed to carry out the theory and simulations presented in this text – hence the title of this text, "Elasticity with Less Stress and Strain".

2. Some of the newer texts of classical elasticity relax the use of symmetric tensors to describe deformation by using the "deformation gradient tensor" instead of symmetric strain tensors, but these authors are also careful to include the many other strain measurements that are symmetric (see Table 16.1). This text uses only the deformation gradient tensor to describe strain.

The problem with trying to capture the deformation with a symmetric matrix is that a symmetric matrix only has six independent values. A Taylor expansion of a general deformation is shown in Eq. 2.24 to contains nine values – the nine values in $\mathcal{F}$. The symmetric matrices $\mathbf{C}$, $\mathbf{B}$, etc., do not capture all of the information about the deformation. In particular, if we decompose $\mathcal{F}$ using Eq. 9.1 we get

$$\mathcal{F} = \mathcal{R}_2 \ \mathcal{L} \ \mathcal{R}_1. \tag{16.1}$$

The matrix C is then

$$\mathbf{C} = \mathcal{F}^T \mathcal{F} = \mathcal{R}_1{}^T \mathcal{L}^T R_2{}^T \mathcal{R}_2 \ \mathcal{L} \mathcal{R}_1 = \mathcal{R}_1{}^T \mathcal{L}^T \mathcal{L} \ \mathcal{R}_1 \tag{16.2}$$

so that the information contained in $\mathcal{R}_2$ is lost. Similarly,

$$\mathbf{B} = \mathcal{F} \ \mathcal{F}^T = \mathcal{R}_2 \ \mathcal{L} \ \mathcal{R}_1 \mathcal{R}_1{}^T \mathcal{L}^T R_2{}^T = \mathcal{R}_2{}^T \mathcal{L}^T \mathcal{L} \ \mathcal{R}_2 \tag{16.3}$$

and the information in $\mathcal{R}_1$ is lost.

This loss of information makes the use of these symmetric tensors to describe deformation in a computer simulation difficult indeed. Using the deformation gradient tensor avoids this problem.

3. Most texts on classical elasticity state that the stress tensor must be symmetric in order that the sum of torques be zero for each node. As a result, even though the anti-symmetric "first Piola-Kirchhoff stress tensor" is sometimes defined in classical texts, it is quickly replaced with the "second Piola-Kirchhoff stress tensor" or the "Cauchy stress tensor", both of which are symmetric stress tensors.

The reason for the assumption that stress tensors be symmetric is argued by many authors, e.g., Spencer[1], pg. 52, provides a convincing "proof". Spencer's argument

Fig. 16.1 The rectangle on
the left will rotate since the
sum of torques (or total
shear stress) is not zero. The
figure on the right is stable,
having the sum of torques
zero (or total shear stress
equal)

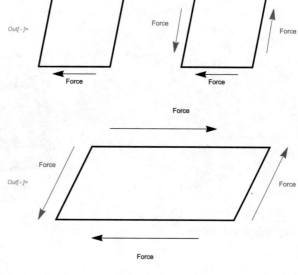

Fig. 16.2 This figure has
equivalent shear stresses,
but not equivalent strains

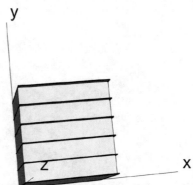

Fig. 16.3 Layers in a cell

boils down to the following: If I apply a force to produce a shear stress in one
direction, I must also apply an equivalent force in the opposite directions in order
that the body remain stationary. (See Fig. 16.1.)

From this it is usually concluded that any stress tensor must be symmetric. The
error in reasoning is that although the final stress must be equal for the body to
remain stationary, this does not mean that the deformation of the body (i.e., the
strain) must be the equal. So, for example, in Fig. 16.2 the stresses are equal, but the
strains are not.

Thus, requiring the final stresses are equal does not require $\frac{\partial \epsilon}{\partial (\partial x_1/\partial X_2)}$ to be equal
to $\frac{\partial \epsilon}{\partial (\partial x_2/\partial X_1)}$ because the deformations $\partial x_1/\partial X_2$ and $\partial x_2/\partial X_1$ do not have to be equal.

In particular, consider a unit cube with layers parallel to the x-axis (Fig. 16.3).
Note that in the x-shear shown in Fig. 16.4, the layers are not stretched. In that

Fig. 16.4 Shear in the x

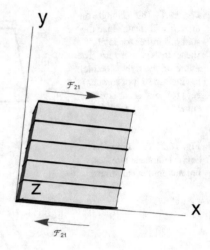

Fig. 16.5 Shear in the y

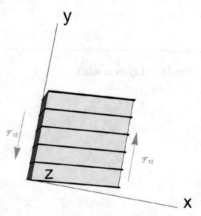

deformation, the side of the cube that was on the y-axis is stretched. The side that was on the x-axis is not stretched. However, in the y-shear shown in Fig. 16.5, the layers are stretched because the side of the cube that was on the x-axis is stretched. So, if the layers are less stretchy than the material holding them together, the force required to produce the deformation in Fig. 16.4 will be less than the force needed to produce the same amount of deformation shown in Fig. 16.5 for the same deformation. This requires $\mathcal{P}_{ij} \neq \mathcal{P}_{ji}$, or in other words, a nonsymmetric stress tensor.

Now in order for each cell in Figs. 16.4 and 16.5 to be stationary, the sum of torques need to be zero. But this would be produced by a different amount of stretch of the two sides of the material. Thus, we would have $\mathcal{P}_{ij} = \frac{\partial \epsilon}{\partial x_i / \partial x_j}$ defined so that $\mathcal{P}_{ij} \neq \mathcal{P}_{ji}$, but yet the final stresses would be equal. This is possible because $\mathcal{P}_{ij}$ and $\mathcal{P}_{ji}$ are evaluated with different amounts of strain. (Note that this also requires that the strain matrix be allowed to be nonsymmetric.)

4. The equation of motion as discussed in classical elasticity is almost always expressed in terms of ρ = mass/(current volume). When this is done, the stress terms must also be expressed in the current volume. But in this text, the equation of motion uses ρ_0 to be mass/(initial volume). This may seem a bit odd, but the result is a simple relationship between force per initial area and energy per initial volume, $\mathcal{P}_{ij} = \frac{\partial \epsilon}{\partial x_i / \partial x_j}$. This is better for computer simulations, because the initial positions of the material are known. If we express the equation of motion in terms of the final positions of the material, we must solve both the equation of motion and solve for the final positions *in* the equations of motion at the same time.

5. Tensor properties are usually introduced early in classical elasticity, but not in this text. Instead stress and strain are expressed only as matrices. This is not to say that stress and strain are not tensors; they are, but as demonstrated in this text, most of tensor properties that distinguish a matrix from a tensor are not needed to describe elasticity in a way that is useful for engineers.

The main differences between a matrix and a tensor are the following three properties: (A) a tensor is a map of one vector field to another, (B) a tensor transforms a particular way if coordinates are rotated, and (C) a tensor may have more than two subscripts (i.e., be of higher order than 2). Only the first property of a tensor is used in this text – and that property is clearly defined in Chap. 2 of this text. With regard to the second property, only one reference coordinate system is used throughout this text. All calculations are done relative to the reference coordinate system, which is never rotated. So, the second property – although true of stress and strain – is not needed. Finally, no tensors are introduced over second order, and therefore the third property is not needed in this text. Since these tensor properties are not needed, the reference to the tensor nature of stress and strain is omitted.

Direct Comparisons with Classical Elasticity

Stress

In order to compare the approach of this text to classical elasticity, I have chosen the classical elasticity text by Spencer[1] to use as a comparison. I find his notation and formulation to be close to that presented in this text. In the following I have used Spencer's notation for his equations to stay true to his text – that is, bold capital letters are second-order tensors, small letters are first-order tensors (vectors), and initial point locations are labeled with a bold capital X and final point locations with a small plain, x. Scalars are plain, not bold. The page numbers and equation numbers referenced in this chapter are from Spencer[1].

The equations of motion are stated in this text in Eqs. 3.16 and 4.41, and ϵ is expressed as a function of $\frac{\partial x_i}{\partial X_j}$. The corresponding equations in Spencer are as follows.

Equation 3.16 is expressed by Spencer in terms of the first Piola-Kirchhoff stress tensor, $\mathbf{\Pi}_{Ri}$ (Spencer: Eq. 9.38, pg. 134).

$$\frac{\partial \mathbf{\Pi}_{Ri}}{\partial \mathbf{x}_R} + \rho_0 \ \mathbf{b}_i = \rho_0 \mathbf{f}_i, \tag{16.4}$$

where

$\mathbf{f}_i =$ acceleration vector (of the mass)
$\mathbf{b}_i =$ gravitational or body force per unit mass
$\rho_0 =$ mass per unit original volume.

Equation 4.41 is expressed by Spencer in terms of the energy per unit original volume, W. (Spencer: unnumbered, first equation on pg. 139):

$$\mathbf{\Pi}_{Ri} = \frac{\partial W}{\partial \mathbf{F}_{iR}}, \tag{16.5}$$

with (Spencer: Eq. 6.18, pg. 69)

$$F_{iR} = \frac{\partial x_i}{\partial x_R} \tag{16.6}$$

and W a function of $\frac{\partial x_i}{\partial X_j}$ is stated as (Spencer: Eq. 10.1, pg. 137)

$$W = W(\mathbf{F}). \tag{16.7}$$

Comparing Eqs. 16.4 to 3.16, 16.5 to 4.41, and the definition of W to ϵ and F to $\mathcal{F}$, we find that we can identify that the transpose of $\mathbf{\Pi}_{Ri}$ is $\mathcal{P}_{ij}$, W is ϵ, and F is $\mathcal{F}$. However, the drive to make stress symmetric and express the equation of motion in terms of the final point positions leads Spencer to express the equation of motion as (Spencer: Eq. 7.22, pg. 98)

$$\frac{\partial \mathbf{T}_{ij}}{\partial x_i} + \rho \ \mathbf{b}_j = \rho \ \mathbf{f}_j \tag{16.8}$$

with T the symmetric Cauchy stress tensor, $\mathbf{b}$ the direction of gravity, and $\mathbf{f}$ the acceleration of the body.

The equation of motion in terms of **T** (Eq. 16.8) appears just as simple as the equation of motion in terms of **Π** (Eq. 16.4). However, this is deceptive because the derivatives in Eq. 16.8 are in terms of x_i, which are the final positions of the points instead of X_i, the initial positions. Also the matrix **T** is symmetric and does not contain all the deformation information (i.e., six independent terms in **T**, whereas **Π** has nine independent terms).

The Invariants

The tensor property of strain is used in classical finite elasticity to define the invariants. That is, the invariants of a symmetric strain matrix are invariant with respect to a rotation of the reference coordinates, i.e., those functions of the components of $C = \mathcal{F}^T \mathcal{F}$ which are unchanged with

$$C' = \mathcal{R} \ C \ \mathcal{R}^T, \tag{16.9}$$

where $\mathcal{R}$ is any rotation matrix.

The resulting invariants as expressed by Spencer[1] (Spencer: Eq. 9.27, pg. 130) are

$$
\begin{aligned}
i_1 &= \text{Trace} \ [C] \\
i_2 &= 1/2 \ (\text{Trace}[C])^2 - 1/2 \ \text{Trace}\left[C^2\right] \\
i_3 &= \text{Determinant} \ [C].
\end{aligned}
\tag{16.10}
$$

In this text, there are two sets of invariants. One set of invariants is invariant to the rotation of the material body as described by

$$\mathcal{F}' = \mathcal{R}_2 \ \mathcal{F} \ \mathcal{R}_1, \tag{16.11}$$

where $\mathcal{R}_1$ and $\mathcal{R}_2$ are any two rotation matrices. These are the invariants which result from an isotropic material being rotated before ($\mathcal{R}_1$) or after ($\mathcal{R}_2$) the deformation ($\mathcal{F}$). The resulting invariants are expressed as $\{\mathcal{I}_1, \mathcal{I}_2, \mathcal{I}_3\}$ or $\{\mathcal{L}_{11}, \mathcal{L}_{22}, \mathcal{L}_{33}\}$ as found in Chaps. 5 and 9. The I_i values are related to the i_i values in Eq. 16.10 as follows:

$$
\begin{aligned}
\mathcal{I}_1 &= i_1 \\
\mathcal{I}_2 &= i_2 \\
\mathcal{I}_3^2 &= i_3.
\end{aligned}
\tag{16.12}
$$

The invariants from Eq. 16.9 are almost identical to the invariants of Eq. 16.11 used in this text, but it is interesting to note here that there are invariants to Eq. 16.9 which are not invariants to Eq. 16.11. These invariants are not appropriate for describing deformations. One example is the trace of the matrix. The trace is invariant to the operations in Eq. 16.9, but not invariant to the operations in Eq. 16.11.

The second set of invariants defined in this text are invariants to the rotation of the material body only after the deformation,

$$\mathcal{F}' = \mathcal{R}\mathcal{F}, \tag{16.13}$$

where $\mathcal{R}$ is any rotation matrix. These invariants are expressed as $\{t_{11}, t_{12}, t_{13}, t_{22}, t_{23}, t_{33}\}$, or $\{I_{11}, I_{12}, I_{13}, I_{22}, I_{23}, I_{33}\}$, for anisotropic bodies as described in Chap. 12. These invariants do not appear in classical elasticity texts.

Direct Comparisons with Rubber Elasticity

Rivlin[5] (pg. 3) chose to describe the deformation of rubber in terms of "an ideal perfectly elastic material", the stored energy of which can be described by the "principal extension ratios" λ_1, λ_2, and λ_3. The three invariants in this text $\{\mathcal{L}_{11}, \mathcal{L}_{22}, \text{and } \mathcal{L}_{33}\}$ are Rivlin's "principal extension ratios". In fact, the experiments in Chap. 10 of this text are essentially the experiments Rivlin and Saunders used to obtain the λ_i values experimentally on a thin rubber sheet (Rivlin[5], pg. 164). To complete the comparison with rubber elasticity as described by Rivlin, Table 16.3 compares the symbols used in Rivlin[5] (pg. 43–45) and those used in this text.

1. Continuum Mechanics, A. J. M. Spencer, Dover Publications, Mineola, NY, 1980.
2. Nonlinear elastic Deformations, R. W. Ogden, Dover Publications, Mineola, NY, 1984.
3. Nonlinear Solid Mechanics for Finite Element Analysis: Statics, Javier Bonet, Antonio J. Gil and Richard D. Wood, Cambridge University Press, Cambridge, UK, 2016.
4. An Introduction to the Theory of Elasticity, R. J. Atkin, N. Fox, Dover Publications, Mineola, NY 1980.
5. Collected Papers of R. S. Rivlin, G.I. Barenblatt and DD Joseph Editors, Springer Science and Business Media, 1997.

Table 16.3 Comparison of symbols in Rivlin[5] to those used in this text

Rivlin		Hardy
ξ	x_1	
η	x_2	
ζ	x_3	
x	X_1	
y	X_2	
z	X_3	
u	$x_1 - X_1$	
v	$x_2 - X_2$	
w	$x_3 - X_3$	
u_x	$\dfrac{\partial x_1}{\partial X_1} - 1$	
u_y	$\dfrac{\partial x_1}{\partial X_2}$	
u_z	$\dfrac{\partial x_1}{\partial X_3}$	
v_x	$\dfrac{\partial x_2}{\partial X_1}$	
v_y	$\dfrac{\partial x_2}{\partial X_2} - 1$	
v_z	$\dfrac{\partial x_2}{\partial X_3}$	
w_x	$\dfrac{\partial x_3}{\partial X_1}$	
w_y	$\dfrac{\partial x_3}{\partial X_2}$	
w_z	$\dfrac{\partial x_3}{\partial X_3} - 1$	
W	ϵ	
$\dfrac{\partial W}{\partial u_x}$	$\mathcal{P}_{11} = \dfrac{\partial \epsilon}{\partial(\partial x_1/\partial X_1)}$	
$\dfrac{\partial W}{\partial u_y}$	$\mathcal{P}_{12} = \dfrac{\partial \epsilon}{\partial(\partial x_1/\partial X_2)}$	
$\dfrac{\partial W}{\partial u_z}$	$\mathcal{P}_{13} = \dfrac{\partial \epsilon}{\partial(\partial x_1/\partial X_3)}$	
$\dfrac{\partial W}{\partial v_x}$	$\mathcal{P}_{21} = \dfrac{\partial \epsilon}{\partial(\partial x_2/\partial X_1)}$	
$\dfrac{\partial W}{\partial v_y}$	$\mathcal{P}_{22} = \dfrac{\partial \epsilon}{\partial(\partial x_2/\partial X_2)}$	
$\dfrac{\partial W}{\partial v_z}$	$\mathcal{P}_{23} = \dfrac{\partial \epsilon}{\partial(\partial x_2/\partial X_3)}$	
$\dfrac{\partial W}{\partial w_x}$	$\mathcal{P}_{31} = \dfrac{\partial \epsilon}{\partial(\partial x_3/\partial X_1)}$	
$\dfrac{\partial W}{\partial w_y}$	$\mathcal{P}_{32} = \dfrac{\partial \epsilon}{\partial(\partial x_3/\partial X_2)}$	
$\dfrac{\partial W}{\partial v_z}$	$\mathcal{P}_{33} = \dfrac{\partial \epsilon}{\partial(\partial x_3/\partial X_3)}$	

Problems

Problem 1 Find the amount of stretch of the layers shown in Fig. 16.5, when the body in Fig. 16.5 is deformed an infinitesimal amount dy, with $\mathcal{F} = \begin{pmatrix} 1 & 0 & 0 \\ dy & 1 & 0 \\ 0 & 0 & 1 \end{pmatrix}$.
Assume the initial cube of material in Fig. 16.5 is $1 \times 1 \times 1$.

Problem 2 Show that the relationships in Eq. 16.12 are true for any $\mathcal{F} =$

$$\begin{pmatrix} f_{11} & f_{12} & f_{13} \\ f_{21} & f_{22} & f_{23} \\ f_{31} & f_{32} & f_{33} \end{pmatrix}.$$

Problem   3 Show   that $\mathcal{F}_1 = \begin{pmatrix} 2 & 0 & 0 \\ 0 & 1 & 0 \\ 0 & 0 & 1 \end{pmatrix}$ and $\mathcal{F}_2 =$

$\begin{pmatrix} 1.87938524 & 0.34202014 & 0 \\ -0.68404028 & 0.939692620 & 0 \\ 0 & 0 & 1 \end{pmatrix}$ have the same right Cauchy-Green deforma-

tion tensor value. If you have a graphics program, plot the results of applying $\mathcal{F}_1$ to a unit cube, and compare it to the results of applying $\mathcal{F}_2$ to the same unit cube.

Problem   4 Show   that $\mathcal{F}_1 = \begin{pmatrix} 2 & 0 & 0 \\ 0 & 1 & 0 \\ 0 & 0 & 1 \end{pmatrix}$ and $\mathcal{F}_2 =$

$\begin{pmatrix} 1.87938524 & 0.68404028 & 0 \\ -0.34202014 & 0.93969262 & 0 \\ 0 & 0 & 1 \end{pmatrix}$ have the same left Cauchy-Green deformation

tensor value. If you have a graphics program, plot the results of applying $\mathcal{F}_1$ to unit cube, and compare it to the results of applying $\mathcal{F}_2$ to the same unit cube.

Problem 5 Show that the invariants of the symmetric matrix $C =$
$\begin{pmatrix} 2 & -.2 & 1.4 \\ -.2 & .3 & .2 \\ 1.4 & .2 & 1.2 \end{pmatrix}$ are unchanged by the rotation of coordinates defined by

Eq. 16.9, with $\mathcal{R} = \begin{pmatrix} 1 & 0 & 0 \\ 0 & \cos\left(20^\circ\right) & -\sin\left(20^\circ\right) \\ 0 & \sin\left(20^\circ\right) & \cos\left(20^\circ\right) \end{pmatrix}$.

Problem 6 Expand Eq. 16.4, and use the expansion to show that Eq. 16.4 is the same as Eq. 3.16 if $\Pi = \mathcal{P}^T$, with $b = \{0, 0, -g\}$ and $\mathbf{f_i} = \{a_1, a_2, a_3\}$.

Problem 7 Write out Eq. 4.45 in terms of Rivlin's notation given in Table 16.3.

Problem 8 Show that $\mathcal{P}_{ij}$ does not equal $\mathcal{P}_{ji}$ with $\mathcal{F} = \begin{pmatrix} 1 & 0 & 0 \\ .1 & 1 & 0 \\ 0 & 0 & 1 \end{pmatrix}$ and for ϵ

defined by Eq. 12.43.

Problem 9 Show that the trace of $\begin{pmatrix} 1.1 & .2 & .4 \\ .2 & .9 & .3 \\ .4 & .3 & 1.2 \end{pmatrix}$ is invariant to the transform in

Eq. 16.9, but not to the transform in Eq. 16.11. Use $\mathcal{R} = \mathcal{R}_1 =$ rotation of 20° about the z-axis and $\mathcal{R}_2 =$ rotation of 30° about the z-axis.

Appendices

Appendix A: Deformation in Jig Coordinates

Chapter 12 stated that the deformation of the material in the fixed coordinate system, $\mathcal{F}_m$, can be transformed into the jig coordinate system using

$$\mathcal{F}_{\text{jig}} = \mathcal{R}^T \mathcal{F}_m \mathcal{R}, \tag{A.1}$$

where $\mathcal{R}$ is rotation which rotates the material in the jig coordinates into the fixed coordinates. This result is derived in this appendix. A 2D illustration will be shown to make the flat page plots easy to see, but all of the steps apply equally well in 3D.

Consider a material placed in service with a layered anisotropy in a fixed coordinate system (top left in Fig. A.1). Consider the jig coordinate system to have been oriented parallel to the layers (top right in Fig. A.1). Using the notation in Chap. 12, $\mathcal{R}$ is the mapping that rotates a body from the jig coordinates system into the fixed coordinate system, so that any dX_{jig} in the jig coordinate system maps to the corresponding dX_m in the fixed coordinate system, i.e.,

$$dX_m = \mathcal{R}\,dX_{\text{jig}}. \tag{A.2}$$

The deformation of the material in the fixed coordinate system is $\mathcal{F}_m$ (bottom left in Fig. A.1),

$$dx_m = \mathcal{F}_m\,dX_m. \tag{A.3}$$

The corresponding deformation in the jig coordinate system is $\mathcal{F}_{\text{jig}}$ (bottom right in Fig. A.1), so that

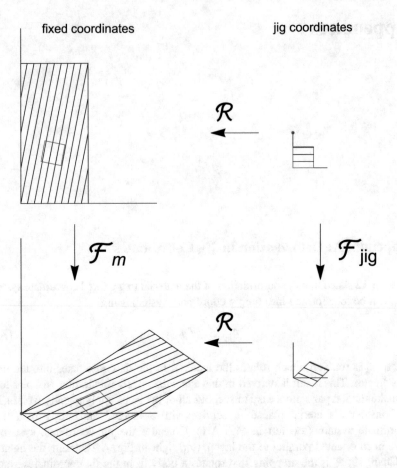

Fig. A.1 A layered material in the top left. A small portion of that material in a jig (top right). Deformation $\mathcal{F}_m$ of layered material in the fixed coordinates (bottom left). Bottom right is the small portion of the material in the jig after $\mathcal{F}_{jig}$ deformation, and $\mathcal{F}_{jig} = \mathcal{R}^T \mathcal{F}_m \mathcal{R}$

$$dx_{\text{jig}} = \mathcal{F}_{\text{jig}} dX_{\text{jig}}. \tag{A.4}$$

The rotation needed to map the final dx_{jig} vector into the material coordinate system is also, $\mathcal{R}$, so we have

$$dx_m = \mathcal{R}\, dx_{\text{jig}}. \tag{A.5}$$

Substitute Eqs. A.2 and A.5 into Eq. A.3

$$\mathcal{R}\, dx_{\text{jig}} = \mathcal{F}_m\, \mathcal{R}\, dX_{\text{jig}}. \tag{A.6}$$

Next multiply both sides of Eq. A.6 by $\mathcal{R}^T$,

Fig. A.2 Left is the small cube of material in the top right of Figure A.1 in the jig coordinate system. Right is the small cube after the deformation $\mathcal{T}$ defined from $\mathcal{F}_{\text{jig}}$ using Eq. 12.21

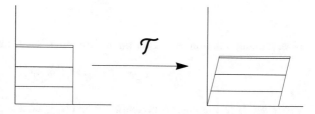

$$dx_{\text{jig}} = \mathcal{R}^T \mathcal{F}_m \, \mathcal{R} \, dX_{\text{jig}} \tag{A.7}$$

and use Eq. A.4 to identify $\mathcal{F}_{\text{jig}}$ as

$$\mathcal{F}_{\text{jig}} = \mathcal{R}^T \mathcal{F}_m \, \mathcal{R}. \tag{A.8}$$

Figure A.1 shows this mapping pictorially.

After $\mathcal{F}_{\text{jig}}$ is found, Eq. 12.21 are used to define $\mathcal{T}$, where $\mathcal{T}$ is the experimental jigdeformation corresponding to the deformation in the fixed coordinates, as shown in Fig. A.2.

Appendix B: Origins of Anisotropic Invariants

This appendix shows the origin of Eq.12.21. This is done by defining an operation which will produce an upper triangularmatrix, $\mathcal{T}$, so that $\mathcal{T}$, is invariant torotations after a deformation. The operation to produce $\mathcal{T}$ is called Gram-Schmidt QRD. After an example calculation of the Gram-Schmidt QRD, Eq. 12.21 are shown to give the same $\mathcal{T}$ as the Gram-Schmidt QRD. In the following, it is assumed that the deformation, $\mathcal{F}$, has already been expressed in jig coordinates (i.e., in this section $\mathcal{F}$ is $\mathcal{F}_{\text{jig}}$.). Finally, a brief explanation of why $\mathcal{F}$ must be in $\mathcal{F}_{\text{jig}}$ instead of $\mathcal{F}_m$ is given.

Why $\mathcal{T}$ Is Invariant Under Rotations After the Deformation $\mathcal{F}$?

There is an operation in linear algebra called QR decomposition. The QR decomposition (also called QR factorization) is a decomposition of any real, square matrix, $\mathcal{F}$, into an orthogonal matrix, Q, and a triangular matrix, $\mathcal{T}$, such that

$$\mathcal{F} = Q\mathcal{T}. \tag{B.1}$$

In the deformation of materials, we do not allow inversions, so Q needs to be replaced with a rotationmatrix, $\mathcal{R}$. Thus, we have

$$\mathcal{F} = \mathcal{R}\,\mathcal{T}. \tag{B.2}$$

Note that any rotation of $\mathcal{F}$ after the deformation, $\mathcal{R}_a\mathcal{F}$, results in

$$\mathcal{R}_a\mathcal{F} = \mathcal{R}_a\mathcal{R}\,\mathcal{T} = \mathcal{R}_1\mathcal{T}, \tag{B.3}$$

where $\mathcal{R}_1$ is just another rotationmatrix and $\mathcal{T}$ is unchanged. Thus, $\mathcal{T}$ is invariant under rotations after any deformation. This is what we need for the energyfunction of an anisotropic material.

Gram-Schmidt QRD

The Gram-Schmidt QR decomposition process (or Gram-Schmidt QRD) is one method to find $\mathcal{T}$ from $\mathcal{F}$ which limits the orthogonal matrix, Q, to a rotational matrix, $\mathcal{R}$.

The steps of Gram-Schmidt QRD are as follows:

1. Extract the column vectors of $\mathcal{F}$ as $\vec{a}$, $\vec{b}$, and $\vec{c}$.
2. Construct a unit vector along $\vec{a}$,

$$\widehat{u}_1 = \frac{\vec{a}}{\sqrt{\vec{a}.{\rightarrow}a}}. \tag{B.4}$$

3. Find the projection of $\vec{b}$ along $\widehat{u}_1$ and subtract this from $\vec{b}$:

$$\vec{g}_1 = \vec{b} - \text{proj}_u \vec{b}. \tag{B.5}$$

Find the unit vector along $\vec{g}_1$:

$$\widehat{u}_2 = \frac{\vec{g}_1}{\sqrt{\vec{g}_1 \cdot \vec{g}_1}}. \tag{B.6}$$

4. Find the projection of $\vec{c}$ along $\widehat{u}_1$ and $\widehat{u}_2$ and subtract both of these from $\vec{c}$:

$$\vec{g}_2 = \vec{c} - \text{proj}_{u_1} \vec{c} - \text{proj}_{u_2} \vec{c}. \tag{B.7}$$

Define a unit vector along $\vec{g}_2$

$$\widehat{u}_3 = \frac{\vec{g}_2}{\sqrt{\vec{g}_2 \cdot \vec{g}_2}}.$$ (B.8)

5. Now the rotationmatrix, $\mathcal{R}$, is the matrix with $\widehat{u}_i$ as the column vectors.

$$\mathcal{R} = (\widehat{u}_1, \widehat{u}_2, \widehat{u}_3)^T.$$ (B.9)

6. The upper triangular matrix, $\mathcal{T}$, is the transpose of $\mathcal{R}$ times $\mathcal{F}$:

$$\mathcal{T} = \mathcal{R}^T \, \mathcal{F}$$ (B.10)

and we have found $\mathcal{R}$ and $\mathcal{T}$ such that

$$\mathcal{F} = \mathcal{R}\mathcal{T}.$$ (B.11)

Example of Gram-Schmidt QRD

For clarity, let us carry out this process on the $\mathcal{F}$ used to deform the fixed coordinates in Fig. B.1.

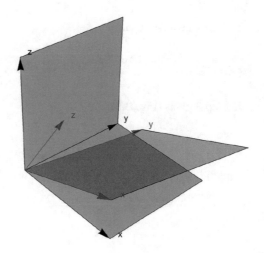

Fig. B.1 Initial coordinate system in black; deformed coordinate system in red. To help see this three-dimensional space on a flat plane, the $z = 0$ and $x = 0$ planes of the original coordinate system are shown in blue, and the $z = 0$ plane of the final coordinate system is shown in red

Fig. B.2 The first new vector, $\widehat{u}_1$, is added to Fig. B.1 and illustrated in green

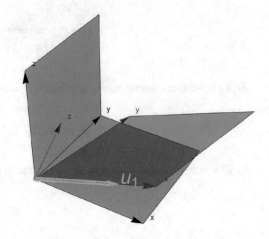

$$\mathcal{F} = \begin{pmatrix} 1.2 & 0.3 & 0 \\ -0.2 & 1 & 0.4 \\ 0.6 & 0.1 & 0.3 \end{pmatrix}. \qquad (B.12)$$

1. The column vectors of $\mathcal{F}$ are as follows:

$$\begin{aligned} \vec{a} &= (1.2, \ -0.2, 0.6) \\ \vec{b} &= (0.3, 1, 0.1) \\ \vec{c} &= (0, 0.4, 0.3). \end{aligned} \qquad (B.13)$$

2. Step 2 gives (Fig. B.2)

$$\widehat{u}_1 = (0.88, -0.15, 0.44). \qquad (B.14)$$

3. Step 3 gives (Fig. B.3).

$$\widehat{u}_2 = \frac{\vec{g}_1}{\sqrt{\vec{g}_1 \cdot \vec{g}_1}} = (0.27, 0.91, 0.09). \qquad (B.15)$$

Fig. B.3 The $\widehat{u}_2$ vector added to Fig. B.2. $\widehat{u}_2$ is a unit vector lying in the plane of the new x-y vectors and perpendicular to $\widehat{u}_1$

4. Step 4 gives (Fig. B.4).

$$\widehat{u}_3 = (-0.44, 0.04, 0.90). \tag{B.16}$$

5. Define $\mathcal{R}$ as the column vectors $(\widehat{u}_1, \widehat{u}_2, \widehat{u}_3)^T$

$$\mathcal{R} = \begin{pmatrix} 0.89 & 0.15 & -0.44 \\ -0.15 & 0.99 & 0.04 \\ 0.44 & 0.03 & 0.90 \end{pmatrix}. \tag{B.17}$$

6. Carry out step 6 to give

$$\mathcal{T} = \begin{pmatrix} 1.36 & 0.16 & 0.07 \\ 0 & 1.04 & 0.40 \\ 0 & 0 & 0.29 \end{pmatrix}. \tag{B.18}$$

In the following section, Mathematica subroutines are used to calculate Gram-Schmidt QRD and Eq. 12.21. Next a general calculation is made showing that these two results are the same for any $\mathcal{F}$. Lastly a comment is made showing why Gram-Schmidt QRD is applied only to $\mathcal{F}_{jig}$ and not directly to $\mathcal{F}_m$. (Fig. B.4).

Fig. B.4 The $\hat{u}_3$ vector is added to Fig. B.3. The $\hat{u}_3$ is a unit vector perpendicular to both $\hat{u}_1$ and $\hat{u}_2$

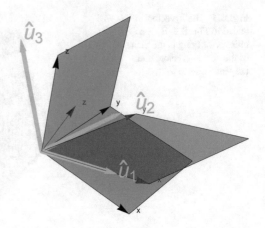

Subroutine for Calculating Gram-Schmidt QRD

```
proj[e_, a_] := Module[{}, (e.a) / (e.e) e]

gramsqrd[amat_] :=
 Module[{a1, a2, a3, u1, e1, u2, e2, u3, e3, qmat, utmat}, tamat = Transpose[amat];
  a1 = tamat[[1]];
  a2 = tamat[[2]];
  a3 = tamat[[3]];
  u1 = a1;
  e1 = u1 / Sqrt[u1.u1];  (* u1 *)
  u2 = a2 - proj[e1, a2];
  e2 = u2 / Sqrt[u2.u2];  (* u2 *)
  u3 = a3 - proj[e1, a3] - proj[e2, a3];
  e3 = u3 / Sqrt[u3.u3];  (* u3 *)
  qmat = Transpose[{e1, e2, e3}];
  utmat = Transpose[qmat].amat;
  {qmat, utmat}]
```

Example calculation.

```
fmat = {{1.2, .3, 0}, {-.2, 1, .4}, {0.6, .1, .3}}
{r, tgrams} = gramsqrd[fmat]
MatrixForm[tgrams]
```

$\{\{1.2, 0.3, 0\}, \{-0.2, 1, 0.4\}, \{0.6, 0.1, 0.3\}\}$

$\{\{\{0.884652, 0.151055, -0.441105\}, \{-0.147442, 0.988149, 0.0426876\},$
$\{0.442326, 0.0272738, 0.89644\}\}, \{\{1.35647, 0.162186, 0.073721\},$
$\{3.46945 \times 10^{-17}, 1.03619, 0.403442\}, \{-1.11022 \times 10^{-16}, 1.11022 \times 10^{-16}, 0.286007\}\}\}$

$$\begin{pmatrix} 1.35647 & 0.162186 & 0.073721 \\ 3.46945 \times 10^{-17} & 1.03619 & 0.403442 \\ -1.11022 \times 10^{-16} & 1.11022 \times 10^{-16} & 0.286007 \end{pmatrix}$$

Subroutine for Calculating Equation 12.21 Values

```
qrd[fmat_] := Module[{a = fmat[[All, 1]], b = fmat[[All, 2]],
    c = fmat[[All, 3]], t11, t12, t13, acb, sacb2, acc, t22, t23, t33},
    t11 = Sqrt[a.a];
         a.b
    t12 = ─── ;
         t11
         a.c
    t13 = ─── ;
         t11
    acb = a × b;
    sacb2 = Sqrt[acb.acb];
    acc = a × c;
         sacb2
    t22 = ───── ;
          t11
          acb.acc
    t23 = ───────── ;
          t11 * sacb2
          Sqrt[(acb.c)²]
    t33 = ───────────── ;
            sacb2
    {t11, t12, t13, t22, t23, t33}]
```

Example calculation:

```
Clear[t11, t12, t13, t21, t22, t23, t31, t32, t33];
fmat = {{1.2, .3, 0}, {-.2, 1, .4}, {0.6, .1, .3}}
MatrixForm[fmat]
{t11, t12, t13, t22, t23, t33} = qrd[fmat]
t12
tmat = {{t11, t12, t13}, {t21, t22, t23}, {t31, t32, t33}} /. t21 → 0 /. t31 → 0 /. t32 → 0
MatrixForm[tmat]
{{1.2, 0.3, 0}, {-0.2, 1, 0.4}, {0.6, 0.1, 0.3}}

⎛ 1.2   0.3   0   ⎞
⎜ -0.2  1     0.4 ⎟
⎝ 0.6   0.1   0.3 ⎠

{1.35647, 0.162186, 0.073721, 1.03619, 0.403442, 0.286007}

0.162186

{{1.35647, 0.162186, 0.073721}, {0, 1.03619, 0.403442}, {0, 0, 0.286007}}

⎛ 1.35647  0.162186  0.073721 ⎞
⎜ 0        1.03619   0.403442 ⎟
⎝ 0        0         0.286007 ⎠
```

General Numerical Proof That Equation 12.21 Provides the Same $\mathcal{T}$ *Matrix as Gram-Schmidt QRD*

```
(* define a general F matrix *)
fmat = {{dx1X1, dx1X2, dx1X3}, {dx2X1, dx2X2, dx2X3}, {dx3X1, dx3X2, dx3X3}}
(* grams result *)
{r, t} = gramsqrd[fmat];
gt11 = t[[1, 1]];
gt12 = t[[1, 2]];
gt13 = t[[1, 3]];
gt22 = t[[2, 2]];
gt23 = t[[2, 3]];
gt33 = t[[3, 3]];
(* Equation 12.21 calculation of the elements of T *)
a = fmat[[All, 1]];
b = fmat[[All, 2]];
c = fmat[[All, 3]];
{t11, t12, t13, t22, t23, t33} = qrd[fmat];
(* show that the two results are the same *)
Simplify[t11 == gt11]
Simplify[t12 == gt12]
Simplify[t13 == gt13]
(* the direct calculation of these last three elements of T was too difficult *)
(*Simplify[t22==gt22];
Simplify[t23==gt23];
Simplify[t33==gt33];*)
(* the following compares the T elements squared *)
Simplify[t22^2 == Simplify[gt22^2]]
(* the t23 element is a bit more difficult for
 Mathematica so it is split it into numerator and denominator *)
Simplify[Numerator[t23] == Numerator[Simplify[gt23]]]
Simplify[Denominator[t23]^2 == Denominator[Simplify[gt23]^2]]
```

The test of the last terms, t33 and gt33, takes a bit longer on my computer:

```
(* save start time *)
start = AbsoluteTime[];
result = Simplify[Numerator[t33^2] == Numerator[Simplify[gt33]^2]];
(* save final time *)
end = AbsoluteTime[];
(* print time elapsed *)
dt = end - start;
Print[result, ":  time (seconds) = ", dt]
```
True: time (seconds) = 25.3563699
```
(* save start time *)
start = AbsoluteTime[];
result = Simplify[Denominator[t33^2] == Denominator[Simplify[gt33]^2]];
(* save final time *)
end = AbsoluteTime[];
(* print time elapsed *)
dt = end - start;
Print[result, ":  time (seconds) = ", dt]
```
True: time (seconds) = 25.0151214

Why Cannot We Just Apply Gram-Schmidt QRD to $\mathcal{F}_m$ Instead of $\mathcal{F}_{jig}$?

Remember that to rotate $\mathcal{F}_m$ to the same coordinate system as $\mathcal{F}_{jig}$, we had to perform

$$\mathcal{F}_{jig} = \mathcal{R}^T \mathcal{F}_m \mathcal{R} \qquad (B.19)$$

But this produces a rotation operation *before* the deformation $\mathcal{F}_{jig}$, and anisotropic deformations should be invariant only for rotations *after* the deformation. Thus, we must first rotate $\mathcal{F}_m$ into $\mathcal{F}_{jig}$ before applying Gram-Schmidt QRD and calculating the upper triangular matrix $\mathcal{T}$.

Appendix C: Euler-Lagrange Equations

Euler-Lagrange Equations in 1D

Consider the function, $f(y, y', x)$, where $y' = \frac{dy}{dx}$. Define the functional J,

$$J = \int_{x_a}^{x_b} f(y, y', x)dx \qquad (C.1)$$

We wish to vary J to find a minimum (or more precisely, an extremum) of J. We wish to do this with fixed x_a and x_b end points. That is, find f such that the variation in J, δJ, is zero:

$$\delta J = 0. \tag{C.2}$$

To accomplish this, define a function $\eta(x)$, which can be any function we choose so long as $\eta(x)$ is zero at the end points, i.e.,

$$\eta(x)|_{x=x_a} = 0 \tag{C.3}$$

$$\eta(x)|_{x=x_b} = 0. \tag{C.4}$$

Since $\eta(x)$ is arbitrary, we can vary y at every point by allowing $y \to y + \alpha\eta(x)$, where α is some unspecified value which is not a function of x. If $y = y + \alpha\eta(x)$,, then $y' = y' + \alpha\eta'(x)$. So, any slight variation of J about any point x gives

$$\frac{dJ(\alpha)}{d\alpha} = \frac{d}{d\alpha}\int_{x_a}^{x_b} f(x, y + \alpha\eta(x), y' + \alpha\eta'(x))dx \tag{C.5}$$

Now set

$$\frac{dJ(\alpha)}{d\alpha} = 0. \tag{C.6}$$

Since α is not a function of x, $\frac{d}{d\alpha}$ passes through the integral, and we have inside of the first integral,

$$\frac{df}{d\alpha} = \frac{\partial f}{\partial y}\frac{\partial y}{\partial \alpha} + \frac{\partial f}{\partial y'}\frac{\partial y'}{\partial \alpha} = \frac{\partial f}{\partial y}\eta + \frac{\partial f}{\partial y'}\eta'. \tag{C.7}$$

Then

$$\frac{dJ(\alpha)}{d\alpha} = \int_{x_a}^{x_b}\left(\frac{\partial f}{\partial y}\eta + \frac{\partial f}{\partial y'}\eta'\right)dx. \tag{C.8}$$

Integrate the second term by parts to give

$$\int_{x_a}^{x_b}\left(\frac{\partial f}{\partial y'}\eta'\right)dx = \frac{\partial f}{\partial y'}\eta\Big|_{x_a}^{x_b} - \int_{x_a}^{x_b}\left(\frac{d}{dx}\left(\frac{\partial f}{\partial y'}\right)\eta\right)dx. \tag{C.9}$$

By Eqs. C.3 and C.4,

$$\left.\frac{\partial f}{\partial y'}\eta\right|_{x_a}^{x_b} = 0. \tag{C.10}$$

and therefore Eq. C.8 reduces to

$$\frac{dJ(\alpha)}{d\alpha} = \int_{x_a}^{x_b}\left(\frac{\partial f}{\partial y} - \frac{d}{dx}\left(\frac{\partial f}{\partial y'}\right)\right)\eta dx \tag{C.11}$$

Since this must be zero for any function, η, the terms in parenthesis must be zero, i.e.,

$$\frac{\partial f}{\partial y} - \frac{d}{dx}\left(\frac{\partial f}{\partial y'}\right) = 0. \tag{C.12}$$

Euler-Lagrange Equations in 2D

Consider the function, $f\left(x, y, \frac{\partial x}{\partial X}, \frac{\partial x}{\partial Y}, \frac{\partial y}{\partial X}, \frac{\partial y}{\partial Y}, X, Y\right)$. Define the functional J,

$$J = \int_{Y_1}^{Y_2}\int_{X_1}^{X_2} f(x, y, \frac{\partial x}{\partial X}, \frac{\partial x}{\partial Y}, \frac{\partial y}{\partial X}, \frac{\partial y}{\partial Y}, X, Y)dXdY. \tag{C.13}$$

We will find the equation for f if

$$\delta J = 0. \tag{C.14}$$

Define $\eta_1(X, Y)$ and $\eta_2(X, Y)$ such that η_1 and η_2 are zero on the boundaries,

$$\eta_1(X, Y)|_{X_a} = 0 \quad \eta_1(x, Y)|_{X_b} = 0 \quad \eta_1(X, Y)|_{Y_a} = 0 \quad \eta_1(x, Y)|_{Y_b} = 0. \tag{C.15}$$

and

$$\eta_2(X, Y)|_{X_a} = 0 \quad \eta_2(x, Y)|_{X_b} = 0 \quad \eta_2(X, Y)|_{Y_a} = 0 \quad \eta_2(x, Y)|_{Y_b} = 0. \tag{C.16}$$

Then define

$$J(\alpha) = \int_{Y_a}^{Y_b}\int_{X_a}^{X_b} f(X, Y, x(X, Y) + \alpha\eta_1, y(X, Y) + \alpha\eta_2, \partial x/\partial X$$
$$+ \alpha\eta_{1X}, \partial x/\partial Y + \alpha\eta_{1Y}, \partial y/\partial X + \alpha\eta_{2X}, \partial y/\partial Y + \alpha\eta_{2Y})dXdY \tag{C.17}$$

$$\frac{dJ}{d\alpha} = \int_{Y_a}^{Y_b} \int_{X_a}^{X_b} \frac{\partial f}{\partial x} \eta_1 + \frac{\partial f}{\partial y} \eta_2 + \frac{\partial f}{\partial (\partial x/\partial X)} \eta_{1X} + \frac{\partial f}{\partial (\partial x/\partial Y)} \eta_{1Y}$$

$$+ \frac{\partial f}{\partial (\partial y/\partial X)} \eta_{2X} + \frac{\partial f}{\partial (\partial y/\partial Y)} \eta_{2Y} dXdY. \tag{C.18}$$

Integration by parts:

$$\int_{X_a}^{X_b} \frac{\partial f}{\partial (\partial y/\partial X)} \eta_{2X} dX = \frac{\partial f}{\partial (\partial y/\partial X)} \eta_2 \bigg|_{X_a}^{X_b} - \int_{X_a}^{X_b} \frac{d}{dX} \left(\frac{\partial f}{\partial (\partial y/\partial X)} \right) \eta_2 dX$$

$$= - \int_{X_a}^{X_b} \frac{d}{dX} \left(\frac{\partial f}{\partial (\partial y/\partial X)} \right) \eta_2 dX \tag{C.19}$$

$$\int_{Y_a}^{Y_b} \frac{\partial f}{\partial (\partial y/\partial Y)} \eta_{2Y} dY = \frac{\partial f}{\partial (\partial y/\partial Y)} \eta_2 \bigg|_{Y_a}^{Y_b} - \int_{Y_a}^{Y_b} \frac{d}{dY} \left(\frac{\partial f}{\partial (\partial y/\partial Y)} \right) \eta_2 dY$$

$$= - \int_{Y_a}^{Y_b} \frac{d}{dY} \left(\frac{\partial f}{\partial (\partial y/\partial Y)} \right) \eta_2 dY \tag{C.20}$$

$$\int_{X_a}^{X_b} \frac{\partial f}{\partial (\partial x/\partial X)} \eta_{1X} dX = \frac{\partial f}{\partial (\partial x/\partial X)} \eta_1 \bigg|_{X_a}^{X_b} - \int_{X_a}^{X_b} \frac{d}{dX} \left(\frac{\partial f}{\partial (\partial x/\partial X)} \right) \eta_1 dX$$

$$= - \int_{X_a}^{X_b} \frac{d}{dX} \left(\frac{\partial f}{\partial (\partial x/\partial X)} \right) \eta_1 dX \tag{C.21}$$

$$\int_{Y_a}^{Y_b} \frac{\partial f}{\partial (\partial x/\partial Y)} \eta_{1Y} dY = \frac{\partial f}{\partial (\partial x/\partial Y)} \eta_1 \bigg|_{Y_a}^{Y_b} - \int_{Y_a}^{Y_b} \frac{d}{dY} \left(\frac{\partial f}{\partial (\partial x/\partial Y)} \right) \eta_1 dY$$

$$= - \int_{Y_a}^{Y_b} \frac{d}{dY} \left(\frac{\partial f}{\partial (\partial x/\partial Y)} \right) \eta_1 dY. \tag{C.22}$$

Using Eqs. C.15 and C.16, we now have

$$\frac{dJ}{d\alpha} = \int_{Y_a}^{Y_b} \int_{X_a}^{X_b} \frac{\partial f}{\partial x} \eta_1 + \frac{\partial f}{\partial y} \eta_2 - \frac{d}{dX} \frac{\partial f}{\partial (\partial x/\partial X)} \eta_1 - \frac{d}{dY} \frac{\partial f}{\partial (\partial x/\partial Y)} \eta_1$$

$$- \frac{d}{dX} \frac{\partial f}{\partial (\partial y/\partial X)} \eta_2 - \frac{d}{dY} \frac{\partial f}{\partial (\partial y/\partial Y)} \eta_2 dXdY \tag{C.23}$$

$$\frac{dJ}{d\alpha} = \int_{Y_a}^{Y_b} \int_{X_a}^{X_b} \left(\frac{\partial f}{\partial x} - \frac{d}{dX} \frac{\partial f}{\partial(\partial x/\partial X)} - \frac{d}{dY} \frac{\partial f}{\partial(\partial x/\partial Y)} \right) \eta_1 dX dY$$

$$+ \int_{Y_a}^{Y_b} \int_{X_a}^{X_b} \left(\frac{\partial f}{\partial y} - \frac{d}{dX} \frac{\partial f}{\partial(\partial y/\partial X)} - \frac{d}{dY} \frac{\partial f}{\partial(\partial y/\partial Y)} \right) \eta_2 dX dY. \qquad (C.24)$$

With η_1 and η_2 arbitrary, then both in parentheses must be zero:

$$\frac{\partial f}{\partial x} - \frac{d}{dX} \frac{\partial f}{\partial(\partial x/\partial X)} - \frac{d}{dY} \frac{\partial f}{\partial(\partial x/\partial Y)} = 0 \qquad (C.25)$$

$$\frac{\partial f}{\partial y} - \frac{d}{dX} \frac{\partial f}{\partial(\partial y/\partial X)} - \frac{d}{dY} \frac{\partial f}{\partial(\partial y/\partial Y)} = 0 \qquad (C.26)$$

Euler-Lagrange Equations with Two Dependent and One Independent Variables

Consider the function $f(x, y, \dot{x}, \dot{y}, t)$.
   Define J as

$$J = \int_0^t L(x, y, \dot{x}, \dot{y}, t) dt. \qquad (C.27)$$

and find L which gives

$$\delta J = 0. \qquad (C.28)$$

Define $\eta_1(x)$ and $\eta_2(y)$ such that at the initial and final times,

$$\eta_1(x)|_t = 0 \quad \eta_1(y)|_t = 0 \quad \eta_1(x)|_{t=0} = 0 \quad \eta_1(y)|_{t=0} = 0 \qquad (C.29)$$

$$J(\alpha) = \int_0^t f(X, Y, x(X, Y) + \alpha\eta_1, y(X, Y) + \alpha\eta_2, \dot{x} + \alpha\dot{\eta}_1, \dot{y} + \alpha\dot{\eta}_2) dt \qquad (C.30)$$

$$\frac{dJ}{d\alpha} = \int_0^t \frac{\partial f}{\partial x}\eta_1 + \frac{\partial f}{\partial y}\eta_2 + \frac{\partial f}{\partial \dot{x}}\dot{\eta}_1 + \frac{\partial f}{\partial \dot{y}}\dot{\eta}_2 dt. \qquad (C.31)$$

Integration by parts:

$$\int_0^t \frac{\partial f}{\partial \dot{y}} \dot{\eta}_2 dt = \frac{\partial f}{\partial \dot{y}} \eta_2 \Big|_0^t - \int_0^t \frac{d}{dt}\left(\frac{\partial f}{\partial \dot{y}}\right) \eta_2 dt = -\int_0^t \frac{d}{dt}\left(\frac{\partial f}{\partial \dot{y}}\right) \eta_2 dt \qquad (C.32)$$

$$\int_0^t \frac{\partial f}{\partial \dot{x}} \dot{\eta}_1 dt = \frac{\partial f}{\partial \dot{x}} \eta_1 \Big|_0^t - \int_0^t \frac{d}{dt}\left(\frac{\partial f}{\partial \dot{x}}\right) \eta_1 dt = -\int_0^t \frac{d}{dt}\left(\frac{\partial f}{\partial \dot{x}}\right) \eta_1 dt. \qquad (C.33)$$

So, we now have

$$\frac{dJ}{d\alpha} = \int_0^t \frac{\partial f}{\partial x} \eta_1 + \frac{\partial f}{\partial y} \eta_2 - \frac{d}{dt}\left(\frac{\partial f}{\partial \dot{x}}\right) \eta_1 - \frac{d}{dt}\left(\frac{\partial f}{\partial \dot{y}}\right) \eta_2 dt \qquad (C.34)$$

$$\frac{dJ}{d\alpha} = \int_0^t \left(\frac{\partial f}{\partial x} - \frac{d}{dt}\left(\frac{\partial f}{\partial \dot{x}}\right)\right) \eta_1 dt + \int_0^t \left(\frac{\partial f}{\partial y} - \frac{d}{dt}\left(\frac{\partial f}{\partial \dot{y}}\right)\right) \eta_2 dt. \qquad (C.35)$$

With η_1 and η_2 arbitrary, then both in parentheses must be zero:

$$\frac{\partial f}{\partial x} - \frac{d}{dt}\left(\frac{\partial f}{\partial \dot{x}}\right) = 0 \qquad (C.36)$$

$$\frac{\partial f}{\partial y} - \frac{d}{dt}\left(\frac{\partial f}{\partial \dot{y}}\right) = 0. \qquad (C.37)$$

Appendix D: Finite Element Example

For the quasi-static examples in the body of this text, Eq. 6.30 (or equivalently the variation of Eq. 14.6 with no gravity or acceleration) is solved by expressing ϵ in terms of the node locations in the simulation. Find minimum is used to then minimize J as in Eq. 6.32.

In this appendix, Eq. 4.45 (or equivalently Eq. 14.12) is solved using a finite element technique with no acceleration and no gravity. This illustrates that either Eq. 6.32 or Eq. 4.45 provides equivalent and useful descriptions of finite elasticity.

The following example is complements of Oliver Rübenkönig of Wolfram Research, makers of Mathematica. The simulation reproduces Fig. 8.18 of the text. It should be run (as all the examples in the book) in Mathematica 13.0 or higher (Fig. D.1contains the code).

Fig. D.1 Initial
configuration of the material
before the deformation

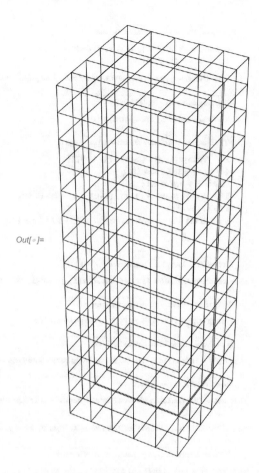

Out[ ]=

Remove all previous definitions:

In[ ]:= `Clear["Global`*"]`

Load the finite element package:

In[ ]:= `Needs["NDSolve`FEM`"]`

Define the amount of stretch:

In[ ]:= `maxpull = 2.0;`

Define size of the material region:

In[ ]:= `xysize = 0.5; zsize = 1.5;`
   `Ω = Cuboid[{-xysize, -xysize, -zsize}, {xysize, xysize, zsize}];`

Create and visualize a first order mesh of the region:

In[ ]:= `mesh = ToElementMesh[Ω, "MaxCellMeasure" → 0.01, "MeshOrder" → 1]; mesh["Wireframe"]`
Out[ ]=

Setup the dependent and independent variables:

```
In[•]:= depVars = {u[x, y, z], v[x, y, z], w[x, y, z]};
       indepVars = {x, y, z};
```

Note that $u = x_1, v = x_2$ and $w = x_3$ and $x = X_1, y = X_2$ and $z = X_3$.

Define the gradient $H_{ij} = \partial x_i / \partial X_j$:

```
In[•]:= H = Grad[depVars, indepVars];
```

Define F in terms of a dummy variable d and the energy ϵ:

```
In[•]:= F = Table[d[i, j], {i, 3}, {j, 1, 3}];
       {a, b, c} = Transpose[F];
       i1 = a.a + b.b + c.c;
       i2 = (a×b).(a×b) + (a×c).(a×c) + (b×c).(b×c);
       i3 = a.(b×c);
       ε = 1.0 (i1 - 3)² + 100 (i3 - 1)² + 2 i1 - 5/4 i2 + 1 i3;
```

Set up a rule that relates the dummy variable d to H:

```
In[•]:= changeVariables = Flatten[Table[d[i, j] → H〚i, j〛, {i, 3}, {j, 3}]];
```

Compute the gradient of the energy ϵ with respect to F:

```
In[•]:= temp = (Grad[ε, #] & /@ F) /. changeVariables;
```

In the final equations set the divergence to zero:

```
In[•]:= eqns = Thread[Inactive[Div][#, indepVars] & /@ temp == {0, 0, 0}];
```

The bottom of the region is fixed:

```
In[•]:= bcs = {DirichletCondition[Thread[depVars == indepVars], z == -zsize]
       }
```

```
Out[•]= {DirichletCondition[{u[x, y, z] == x, v[x, y, z] == y, w[x, y, z] == z}, z == -1.5]}
```

Define rotation matrix to define rotation of end of material:

```
In[•]:= rotmat = {{Cos[thy], 0, Sin[thy]}, {0, 1, 0}, {-Sin[thy], 0, Cos[thy]}}.
       {{Cos[th], -Sin[th], 0}, {Sin[th], Cos[th], 0}, {0, 0, 1}} /. th → 80 * π / 180 /.
       thy → -55 * π / 180; deform = ((rotmat).{x, y, maxpull}) ;
```

Define the number of steps to divide the deformation into:

```
In[•]:= nsteps = 200;
```

The initial position of each node to the undeformed mesh coordinates:

```
In[•]:= displacements = ElementMeshInterpolation[mesh, #] & /@ Transpose[mesh["Coordinates"]];
```

Solve the equations with boundary conditions. Use the previous point locations as the starting values to search for the next solution, i.e., "InitialSeeding" = previous point values (Fig. D.2).

Fig. D.2 Final
configuration of the material
shown in red. Initial
configuration of the material
before the deformation
shown in black

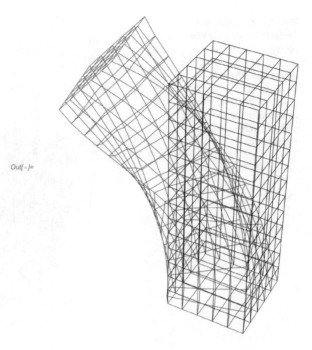

Out[ ]=

Solve the equations:

```
In[ ]:= Do[
    (* compute the pull in each step *)
    {xpull, ypull, zpull} = (deform - {x, y, z}) * step / nsteps + {x, y, z};

    (* implement the pull as a boundary condition *)
    bcs2 =
     Join[bcs, {DirichletCondition[Thread[depVars == {xpull, ypull, zpull}], z == zsize]}];

    (* solve the actual equations *)
    displacements = NDSolveValue[{eqns, bcs2}, depVars[[All, 0]], indepVars ∈ mesh,
      "InitialSeeding" → Thread[Equal[depVars, Through[displacements[x, y, z]]]]
     ];

    PrintTemporary["step = ", step];

    (* create a new, deformed mesh at each step *)
    valuesOnGrid = #["ValuesOnGrid"] & /@ displacements;
    newMesh = ToElementMesh["Coordinates" → Transpose[valuesOnGrid],
      "MeshElements" → mesh["MeshElements"]];

    (* save the mesh at each step *)
    saveMesh[step] = newMesh
    ,
    {step, 1, nsteps}]
```

Fig. D.3 Configuration of the material after 60 steps shown in red. Initial configuration of the material before the deformation shown in black. Also shown are red, green, and blue lines, which show the x, y, and z coordinate axes, respectively

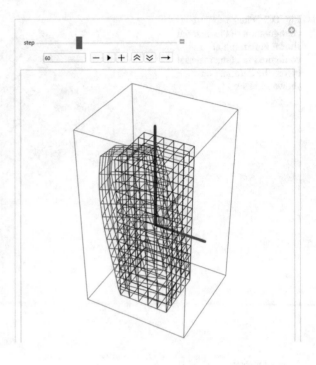

Display the final mesh in red with the original mesh in black:

```
In[ ]:= Show[
        mesh["Wireframe"],
        saveMesh[nsteps]["Wireframe"["MeshElementStyle" → Directive[EdgeForm[Red]]]]]
Out[ ]=
```

The following allows inspection of the results of the simulation for each step in the simulation. If this appendix is put into a Mathematica notebook and run with Mathematica 13.0 or higher, the figure below will be "live", and moving the slider next to the name "steps" will show the results of each step in this simulation. The figure above shows the final configuration of the material. The figure below shows the configuration at step number 60 (Fig. D.3).

Display the results of each step as a "movie":

```
pltcoords = Graphics3D[{Thickness[0.01], Red, Line[{{0, 0, 0}, {1, 0, 0}}],
    Green, Line[{{0, 0, 0}, {0, 1, 0}}], Blue, Line[{{0, 0, 0}, {0, 0, 2}}]}];
Manipulate[Show[{pltcoords,
    mesh["Wireframe"],
    saveMesh[step]["Wireframe"["MeshElementStyle" → Directive[EdgeForm[Red]]]]}],
  {step, 1, nsteps, 1}]
```

Appendix E: Project Ideas

This appendix lists a set of suggested projects for each chapter. These are a bit more involved than the problems at the end of each chapter and may be used for class assignments or for undergraduate or master's thesis projects depending upon the individual project.

Chapter 2

1. The matrix maps in this chapter are second-order tensors. Review the literature on tensors and explain why this is the case. For a bit more, figure out if they are covariant or contravariant tensors and why.
2. Draw a grid on a flat piece of Play-Doh or Silly Putty. Then pull on the edges of the piece so that the grid is distorted. Try to find a map that will match the distorted grid starting with a regular one (see Fig. 2.12).
3. What might the mapping equations look like if you kept the first three terms of a Taylor expansion instead of only the first two shown in Eq. 2.17?

Chapter 3

4. Take a heavy block and put a hook in the center of one side of the block. Place the block on a flat surface. Attach a spring scale to the block and pull at some angle. Do not pull so hard that the block slides or tips over. Note the force magnitude from the spring scale. Measure the angle that the spring scale makes with the sides of the block, and calculate the components of this force. Find the normal force and the shear force on the block. If you want a bit more, pull on the block until it starts to slide or tip over. Then calculate the force (magnitude and direction) of the force needed to keep the block stationary. After you have the result, test your answer by attaching another spring to the block. Does it matter where you attach the new spring? Why?
5. Expand the equation of motion (Eq. 3.16) to include a distribution of charge on the body when placed in an electric field. For more, a dipole of charge in the electric field can be included.
6. Expand the equation of motion (Eq. 3.16) to include a dielectric material in an electric field, for more, a diamagnetic material in a magnetic field.

Chapter 4

7. Take two springs and attach two masses to one spring. Attach the other spring to one of the masses and to a fixed point as shown in Fig. 4.3. Pull on the mass farthest from the fixed point, and describe what you see the middle mass doing. Why? If at first the motion does not seem "interesting", pull faster. For a bit more, if you have good computer skills, write a computer code that will mimic the motion of the center mass.

8. Purchase some Dragon skin® 10 Very Fast, Platinum Silicone – Very Fast Cure, sold by Smooth-On, Inc., 5600 Lower Macungie Road, Macungie, PA 18062, phone 610-252-5800, www.smooth-on.com, material. Make a small rectangular mold out of cardboard or wood (or 3D printer). Pour the material in the mold and let it set. Remove the material from the mold. Glue a piece of wood to opposite sides and pull. What do you see? Why does not the result look like Fig. 4.5? For more, shear the material as in Fig. 4.6 and explain what you see.

Chapter 5

9. Write computer code to rotate a square (or cube) about different angles and different displacements, and display the results. If you have access to an interactive interface, build an interface so that someone can change the angles of rotation and displacements "on the fly".

10. Build an interactive computer display to allow the user to rotate a square, cube, happy face, etc., so that the user can apply rotations before and after a deformation defined by $\mathcal{F}$.

Chapter 6

11. Write a conjugate gradient code to minimize J (Eq. 6.30). For a bit more, use a different optimization scheme than conjugate gradient.

12. If you are not familiar with the calculus of variations, do a report on what it is and how it applies to the equations of deformation (see Chap. 14 as a starting point).

13. Research the literature to find if there is a better method than the conjugate gradient method to find minimums of multidimensional functions.

Chapter 7

14. Write a user interface for the code described in Chap. 7.

15. Add a significant amount of gravity to any of the example problems in Chap. 8.

16. Download an interesting shape from the Internet. Import it into Mathematica (or other) software and watch it "collapse" as gravity is added to it.

17. FindMinimum cannot use the Conjugate Gradient option to solve Lagrange multiplier problems. Find an algorithm that will find "the nearest" or "close" minimum using Lagrange multipliers. For a bit more, include it in the Mathematica code.

18. In each of the examples, the multiple of $(\mathcal{I}_3 - 1)^2$ is set to 100 (large compared the multiples of the other terms). Vary this, and study the relationship between this multiple and Poisson's ratio (the ratio of the width to the compression or extension). Discover for which multiples the software "breaks down".

Chapter 8

19. Use the single grid study, and plot Poisson's ratio (the ratio of the width to the compression or extension).

20. Change the energy equation used in the simulations from the examples given. In each new equation, run the four-point simulation and the cylindrical simulation, and compare the force-compression plots as was done in Figs. 8.11 and 8.12.

21. Use the cylindrical example, but instead of displacing all the points on the top uniformly, press only on a hand full of nodes on the top, and observe the deformation that results. For more, you can try shearing it, twisting it, etc. Each time plot the results.

22. The shape of the stress-strain curve is similar to many biological soft tissues. Research the stress-strain curves of aneurysms, and find an energyfunction which will model them.

23. Plot the force versus deformation for the cylinder in section "Real compression", and compare it to the force versus deformation of a single grid, as in Fig. 8.11.

24. Reproduce the experiment and computer simulation shown in Fig. 8.34.

25. Extrude Play-Doh or cake icing though different "tips", and match the shape of the extruded material using the computer code. For more, create a 3D version of the software to match the results in 3D.

26. Use a water drop from a faucet or a filled water balloon to create the "tear drop" shape shown in the computer simulation in Fig. 8.35. For a bit more, take measurements on the balloon or water drop along with its weight, and find parameters a, b, and c to match these measurements. To *really* do this, you will need to make the code 3D.

27. For any example given in Chap. 8, translate the Mathematica code into a computer language of your choice (e.g., MatLab, Python, C++, Fortran, or other).

28. Push, bend, or distort Play-Doh or clay in some way. Keep track of the deformations you applied, and use the Mathematica code to reproduce the shape (like in Figs. 8.18 and 2.1). If you want a bit more, measure the forces you used to deform the material, and see if you can adjust the energy equation to match both the deformation and the forces applied.

29. The shape of the stress-strain curve in Fig. 8.12 is called a "J-shaped curve". Do a literature search and find materials that behave in this fashion. For more, pick one of the materials, and find an energy function that will match the stress-strain curve of that material.

30. Repeat the simulation of Fig. 2.1 using a plastic deformation simulation as in section "Extrusion study".

Chapter 9

31. Why are the invariants of $\mathcal{F}$ the same as the tensor invariants of $\mathcal{F}^T\mathcal{F}$?

32. Why is the rotationmatrix in 9.7 constructed using the row vectors instead of the column vectors?

Chapter 10

33. Reproduce the experiments in Chap. 10. For more, improve upon the experimental design or technique to find the best fit to energy.

34. Is it possible to use other experiments to find the energy function – see Rivlin (Collected Papers of R. S. Rivlin, Volume I, Edited by G. I. Barenblatt and D. D. Joseph, Springer, 1997, pg 157-194) and Treloar (The Physics of Rubber Elasticity, by L. R. G. Treloar, Oxford Classic Texts, Third Edition, Clarendon Press, Oxford 2005, pg 80-100) for some ideas.

35. Design and carry out a better experiment than the one given in the text to determine if Dragon Skin is incompressible or not.

Chapter 11

36. In Chap. 11, friction was added to the code as a force proportional to the velocity of the nodes. Introduce friction into the model by the relative velocity of the nodes (i.e., add the internal friction of the material to the model).

37. Repeat the oscillating experiment of Chap. 11 but make the initial displacement a shear displacement instead of a compression. Plot the results for different amounts of friction. For more, compare the oscillations to that of a cube of jello – either figuratively or numerically.

Chapter 12

38. Repeat the calculation on the spring model of a material (Eq. 12.23) to obtain ϵ but use different spring constants for the different springs.
39. Repeat the second numerical example in Chap. 12 but use thin elastic sheets instead of fibers.
40. The procedure described here requires that the friction on the surface of the jig be minimal. This may be difficult in practice. As an alternative, it is possible to map the energyfunction of an anisotropic material by cutting pieces of the material with different orientations of the anisotropy and using the procedure described in Chap. 10. Describe mathematically how this would work. For a bit more, perform the experiments of Chap. 10 to find the energy function of an anisotropic material (e.g., construct the material by embedding thin elastic thread into the material when it is formed).

Chapter 13

41. Incorporate the force, displacement, and internal force plots into the 3D core computer experiments of Chap. 8.
42. Compare the shape differences in the top left figure and the bottom right figure in Fig. 13.16 to shapes created by 3D core compressions as in Chap. 8.
43. Restate the differential Eq. (4.45) in terms of finite differences. For more, solve these equations for one or more of the examples in the text, and compare the results.

Chapter 14

44. Recast the Euler-Lagrange approach here into a Hamiltonian. Write out the Hamiltonian equations in terms of a conical momentum.
45. Build a 3D core model with fracturing (or modify the code in Chap. 8 to include fracturing).

Chapter 15

46. From Eq. 15.51 try many different values of A through H, and plot the force versus compression curves to explore how each parameter affects the plot (Fig. 8.2).
47. Use the linear elasticityenergyfunction, Eq. 15.22, in the computer code to create various deformations. Some strange forms can result (see Finite Element

Method simulation of 3D Deformable Solids, by Eftychios Sifakis and Jernej Barbic, Morgan & Claypool Publishers, 2016.)

48. Find the relationship between T and Π. Show that the relationship given by Rivlin (Collected Papers of R. S. Rivlin, Volume I, Edited by G. I. Barenblatt and D. D. Joseph, Springer, 1997, Equation 7.5, pg 35) and Spencer (Continuum Mechanics by A. J. Spencer, Dover Publications, Mineola, NY, 1980, equation 9.36, pg 133) is the same.

49. Find other combinations of A and H that would give a linear stress-strain response, and define these in terms of Y and ν.

Chapter 16

50. Restate the 12 equations of the 4 nodes bounding a tetrahedron as 3 equations to fix the center of mass, 3 equations to fix the rotation about the center of mass, and 6 equations defining $\mathcal{T}_{ij}$.

Index

A

Acceleration, 9, 13, 14, 33, 38, 41, 42, 62, 63, 68, 72, 76, 147, 158, 164, 208, 209, 236, 258

Affine map, 18, 20–22, 25, 27, 31

Almasi strain, 230

Anisotropic material, 12, 55, 165–185, 246, 267

B

Bending, 89, 90, 102–109

Biot strain, 230

Biot stress, 231

Body force, 33, 34, 38, 39, 53, 55, 207, 209, 236

Boundary force, 79, 155–157, 159, 187, 196, 208, 210

C

Cauchy-Green deformation tensor, 240

Cauchy stress, 211, 231, 232, 236

Cauchy stress tensor, 231, 232, 236

Classical elasticity, 229, 230, 232–238

Compression, 41, 58, 89–101, 122, 125, 126, 130, 131, 134, 138, 143, 157, 159, 160, 164, 181, 182, 188–192, 195, 197–201, 265–267

Connectivity, 77

Coordinates, 10, 17, 19–21, 26–30, 33, 35, 40, 41, 53, 56, 58, 59, 62, 76, 80, 83, 87, 120, 122, 126, 157, 165–167, 170, 172, 174, 177, 178, 184, 185, 206, 229, 235, 237, 241, 243–245, 247, 253, 262

Curvature, 89–91, 103–107, 117

D

Deformation, 12, 17–31, 33, 34, 40, 41, 43, 50, 51, 53, 55, 57–63, 67–70, 72, 73, 76, 77, 81–83, 85, 87, 89–92, 94, 97, 100, 102, 103, 107–111, 114–117, 119, 120, 122–124, 126, 127, 129, 131, 132, 134, 137, 147, 148, 157, 163–166, 168, 174, 176–179, 181, 184, 185, 187–192, 194, 196, 198, 199, 201, 203, 205–209, 211–213, 215–217, 221, 222, 224–227, 229, 230, 232–234, 237, 238, 240, 243–246, 253, 259, 261, 262, 264–267

Deformation gradient, 26, 58, 91, 147, 215, 230, 232, 264, 265

Density, 39–41, 68, 72, 76, 77, 164, 207

Determinant, 4, 120

Dirichlet boundary conditions, 64, 77, 208

Discrete model, 67

Displacement, 21, 44–46, 50, 63, 76, 91, 97, 108, 112, 133, 148, 164, 165, 168, 170, 187, 188, 191, 194–197, 204, 221, 230, 264, 266, 267

E

Eigenvalues, 120, 121

Eigenvectors, 121

Einstein notation, 3–6, 24, 39, 40

Energy function, 60, 90–92, 94, 97–99, 114, 129, 142–144, 165, 167, 177, 179, 180, 182, 184, 193, 194, 198, 203, 204, 221, 246, 265–267

Energy of deformation, 43, 50, 51, 53, 63, 67–70, 72, 73, 76, 81–83, 87, 91, 97, 100, 102, 103, 107, 108, 119, 122, 132,

Printed in the United States
by Baker & Taylor Publisher Services